SOUTHEAST ASIAN CULTURE AND ~~HERITAGE~~
IN A GLOBALISING WORLD

T0249930

Heritage, Culture and Identity

Series Editor: Brian Graham,
School of Environmental Sciences, University of Ulster, UK

Other titles in this series

Southeast Asian Culture and Heritage in a Globalising World

Diverging Identities in a Dynamic Region

Edited by

RAHIL ISMAIL
National Institute of Education,
Nanyang Technological University, Singapore

BRIAN J. SHAW
The University of Western Australia

OOI GIOK LING
National Institute of Education,
Nanyang Technological University, Singapore

Routledge
Taylor & Francis Group

LONDON AND NEW YORK

First published 2009 by Ashgate Publishing

Published 2016 by Routledge
2 Park Square, Milton Park, Abingdon, Oxfordshire OX14 4RN
711 Third Avenue, New York, NY 10017, USA

First issued in paperback 2016

Routledge is an imprint of the Taylor & Francis Group, an informa business

Copyright © Rahil Ismail, Brian J. Shaw and Ooi Giok Ling 2009

Rahil Ismail, Brian J. Shaw and Ooi Giok Ling have asserted their right under the Copyright, Designs and Patents Act, 1988, to be identified as the editors of this work.

All rights reserved. No part of this book may be reprinted or reproduced or utilised in any form or by any electronic, mechanical, or other means, now known or hereafter invented, including photocopying and recording, or in any information storage or retrieval system, without permission in writing from the publishers.

Notice:
Product or corporate names may be trademarks or registered trademarks, and are used only for identification and explanation without intent to infringe.

British Library Cataloguing in Publication Data
Southeast Asian culture and heritage in a globalising world :
    diverging identities in a dynamic region. - (Heritage,
    culture and identity)
    1. Ethnology - Southeast Asia 2. Globalization - Southeast
    Asia 3. Southeast Asia - Civilization
    I. Ismail, Rahil II. Shaw, Brian J. III. Ooi, Giok Ling
    959

Library of Congress Cataloging-in-Publication Data
Southeast Asian culture and heritage in a globalising world : diverging
identities in a dynamic region / edited by Rahil Ismail, Brian Shaw, and Ooi
Giok Ling.
    p. cm. -- (Heritage, culture, and identity)
    ISBN 978-0-7546-7261-6
    1. Southeast Asia--Civilization. 2. Ethnology--Southeast Asia. 3.
Globalization--Southeast Asia. I. Ismail, Rahil. II. Shaw, Brian J. III. Ooi,
Giok Ling.

    DS523.2.S53 2008
    959--dc22

                                                                    2008032627

ISBN 13: 978-1-138-27092-3 (pbk)
ISBN 13: 978-0-7546-7261-6 (hbk)

Contents

List of Figures and Tables

Figures

Tables

List of Contributors

Mark Baildon is an Assistant Professor in Humanities and Social Studies Education at the National Institute of Education (Nanyang Technological University, Singapore). He has a PhD in Curriculum, Teaching, and Educational Policy from Michigan State University, a Masters in Social Sciences from Syracuse University, and a B.A. in history and psychology from the University of Rochester. His teaching and research interests include inquiry-based and critical social studies education, the uses of technology to support disciplined inquiry practices, multiliteracies, and teacher learning. Mark has also taught social studies in secondary schools in the United States, Israel, Singapore, Saudi Arabia, and Taiwan. Recent publications include, *Negotiating epistemological tensions in thinking and practice: A case study of a literacy and inquiry tool as a mediator of professional conversation* (with J. Damico, under review) and *Examining ways readers engage with Web sites during think aloud sessions* (with J. Damico).

Kevin Blackburn is currently an Associate Professor in History, National Institute of Education, Nanyang Technological University, Singapore. Since 2001, he has also presented various interviews for Radio Australia, ABC Radio, and talkback radio stations, Mediacorp's CNA and Chinese Channel 8, CNN, and newspapers, *Straits Times*, *Lianhe Zaobao*, *Shin Min Daily News* on the Japanese Occupation. He is also a referee and reviewer for the journal, *Australian Studies* (the British Australian Studies Association, King's College, University of London). His most recent publications include *Forgotten Captives in Japanese Occupied Asia: National Memories and Forgotten Captivities*, London, Routledge (with Karl Hack).

Nancy Hudson-Rodd was formerly Senior Lecturer and is now Adjunct Associate Professor in the School of International, Cultural and Community Studies Mt Lawley Campus Edith Cowan University, Western Australia.

Rahil Ismail is currently an Assistant Professor in National Institute of Education, Nanyang Technological University, Singapore in the Humanities and Social Sciences Education Academic Group. She has expertise in the areas of multicultural studies and education with specific reference to Singapore, and has acted as consultant and facilitator for community organisations, such as *People's Association* and *Central Singapore Joint Social Service Centre*. This is intertwined with her other research interests in heritage studies and international relations. Her publications encompass this wide range of interests as with her teaching duties which included teaching and coordinating American history and politics, international relations, multicultural studies, film history, the Vietnam War and conflict and cooperation. Recent publications include 'Children's Experiences of Multiracial Relationships in Informal Primary School Settings' (co-author), 'Singapore's Malay-Muslim Minority: Social Identification in a Post-9/11 World' (with Brian J. Shaw), 'Ignoring the Elephant in the Room: Racism in the War on Terror', 'Ramadan and Bussorah Street: The Spirit of Place', 'Ethnoscapes, Entertainment and 'Eritage in the Global City: Segmented Spaces in Singapore's Joo Chiat Road' (with Brian J. Shaw).

Ooi Giok Ling is currently Professor, National Institute of Education, Nanyang Technological University. Previously she was a Senior Research Fellow at the Institute of Policy Studies, Singapore; Adjunct Associate Professor Department of Geography, National University of Singapore; and has served as Director, Research Division, at the Ministry of Home Affairs in Singapore. She has published more than eighty refereed articles and chapters in books and has authored or co-authored twelve books, most recently *Civic Space and the Rise of Civil Society in a Globalising World*, London: Routledge (forthcoming); *Changing Geographies and Global Issues of the 21st Century* (2006) Singapore: Pearson/Prentice-Hall; *Sustainability and Cities – Concept and Assessment* (2005) World Scientific Press, Singapore; *Housing in Southeast Asian Capital Cities* (2005) Southeast Asian Book Series, Institute of Southeast Asian Studies, Singapore, and *The Future of Space – Planning, Space and the City* (2004) Eastern Universities Press, Singapore.

Brian J. Shaw is Senior Lecturer in the School of Earth and Geographical Sciences at The University of Western Australia, Perth. His research into urban development, heritage and tourism issues has been widely published in journals such as *Australian Geographer, Current Issues in Tourism, GeoJournal, International Journal of Heritage Studies, Malaysian Journal of Tropical Geography* and *Urban Policy and Research*. His recent books include joint authorship of *Beyond the Port City: Development and Identity in C21st Singapore* (Pearson 2004) and co-editorship of *Challenging Sustainability: Urban Development and Change in Southeast Asia* (Marshall Cavendish 2005).

Michael Theno was formerly an Assistant Professor in Menlo College, California. He has 17 years teaching experience at the undergraduate and graduate levels with students diverse in age, ability, nationality and ethnicity. His areas of teaching competence include Political Science, Humanities, Organizational Behavior and Development, Diversity, Management, and Public Administration. This extends into developing course curricula as well as executing other activities beyond traditional lectures. His research interests also include the political and economic development of Indochina. His most recent publications include *The Lao Hmong of Watt ham Krabok: A Moment of Enactable Policy* (2006) with M. Speck.

Ambar Widiastuti graduated cum laude from the Faculty of Social and Political Sciences, Gadjah Mada University, Yogyakarta, Indonesia where she achieved 'Best Graduate from Department of International Relations' with a thesis entitled 'The Role of Education Policies in Managing Racial Harmony in Singapore' (2006). She is currently working with the World Health Organization (WHO) in Yogyakarta, Indonesia and was formerly a tutor at the Muhammadiyah University of Yogyakarta.

Johannes Widodo is an Associate Professor at the Department of Architecture with a joint appointment in Asian Cities Cluster of Asia Research Institute (ARI) at the National University of Singapore. He received his PhD in Architecture from the University of Tokyo, Japan (1996), Master of Architectural Engineering degree from Katholieke Universiteit Leuven, Belgium (1988), and his first professional degree in Architectural Engineering (Ir.) from Parahyangan Catholic University in Bandung, Indonesia (1984). His area of specialisation includes architecture, urban history and morphology of Southeast Asian cities, Asian modernity, and Heritage Conservation. His current on-going research project is on the morphology and transformation of the coastal cities in Indonesia, Malaysia and Thailand, funded by the National University of Singapore (2005–2008). He wrote *The Boat and the City – Chinese Diaspora and the Architecture of Southeast Asian Coastal Cities*, published by Marshall Cavendish Academic, Singapore (2004). Recently he contributed two chapters on Modern Indonesian Architecture and on the Chinese Diaspora Architecture in *The Past in the Present – Architecture in Indonesia*, edited by Peter J.M. Nas, published by NAi publisers, Rotterdam (2006). He is the editor of *ARCASIA Architectural Timeline Chart* book, published by the Architects Regional Council Asia and University of Santo Tomas Publishing House, Manila (2006).

Foreword

Southeast Asian Culture and Heritage in a Globalising World is a welcome scholarly addition to the ongoing conversation about global futures, especially as it pertains to Southeast Asia. This volume, in papers that look both backward and forward, is especially welcome in that the contributors are insiders and those with an intimate knowledge of the region. The voices are therefore authentic and the analysis both rigorous and sympathetic.

Southeast Asia's ancient and recent histories, its diversity and its mix of future and past in its urban, and still considerable rural habitats, are unique; it is the crossroad of metropolitan and regional cultures. Southeast Asia is simultaneously coming to terms with persistent tradition, modernity and post modernity. Its success and failures in managing wrenching change will offer valuable insights into how change processes involving the local, national, regional and global can be managed.

Of particular interest is the serious attention devoted in this volume to the ways in which traditional resources or heritage is used, deliberately and accidentally, worked and reworked to satisfy multiple audiences. 'Instant Asia' may be a catchy marketing slogan but it grossly undervalues enduring Asia. Several papers in this volume look at several aspects ranging from curriculum reform, ethnic enclaves, tourism islands, and commemorative spaces, using them as illustrative ethnoscapes to detail the ways in which change is being confronted and managed. One concern is the possibility that the new cultural geographies being created by change may not be sustainable or provide for equitable and sustainable development. That remains to be seen but I remain confident in the resilience of enduring values and ways of living.

Professors Ismail, Shaw and Ooi are to be congratulated on their efforts in turning conference papers into a well-edited and compelling volume. I am certain that it will be a major text in university courses and indeed read more widely amongst those who will want to better understand the region.

S. Gopinathan, Professor and Head
Centre for Research in Pedagogy and Practice
National Institute of Education
Nanyang Technological University
Singapore

Preface

The Southeast Asian Geography Association (SEAGA) is proud to have provided the platform upon which the ideas for this book of insightful essays on heritage and identity issues in the region have been developed. The association is an international network of scholars, academics, educators and professionals, who are working on, as well as in, the Southeast Asian region. This network has met biennially in different Southeast Asian locations since its 1990 inception in Brunei Darussalam. Its success is a tribute to the vision displayed by Professor Goh Kim Chuan, the early driving force behind the Association, and more latterly Professor Ooi Giok Ling who has presided over the last two SEAGA meetings from 2006.

During the 2006 SEAGA International Conference held in Singapore, a number of the authors of the essays in this book met to deliberate the politics of heritage, culture and identity in the port cities as well as other coastal cities in Southeast Asia, conservation of ethnic neighbourhoods in fast growing world cities as well as the meanings of regional and local identities in a globalising world. These are themes that are important to the fast growing and changing region of Southeast Asia and its people. While material needs – food, housing, energy, infrastructure for health, transport and environment among others – remain important in shaping the cultural landscapes of the region, the turmoil that continues to challenge state and society in the Southeast Asian region appears to revolve around the nation-state and its meanings in relation to ethnicity, cultural heritage, place and belonging.

The essays in the book have noted that with global competition, governments in Southeast Asia have been responding in myriad ways in the bid to attract international investors, businesses and tourists. Contestation between local and global needs have emerged throughout the region as governments decide between investments in international airports and telecommunications infrastructure or basic housing, clean water supply and such more localised needs. The bid by national and city governments to integrate more closely with the global economy also implies rapid change that has led to social fragmentation and the exclusion of large segments of society from the benefits that globalisation purportedly brings.

In the process of change, the authors rightly point out that the state, market and institutions in society in the Southeast Asian region act as cultural and social gatekeepers. Much of the time, policy decisions and market developments will have impact and implications for the context in which identities are formed and shaped together with its attendant meanings. Although globalisation suggests that de-territorialisation will be a major outcome in processes that are changing the region, clearly trends point to the contrary as varying forms of civic engagement appear to be rallying around identities that are very much linked to nationalities, ethnicity and common cultural backgrounds. These civic processes involving citizens demanding political reforms and attention to neglected social policies have been organised around places that have become icons of political reforms and change – Edsa in Manila in the Philippines and Independence Square in Bangkok, Thailand.

Southeast Asia has always been at the crossroads of cultural exchange and the meeting of varying cultures – east and west, Asian, Southeast and East Asian. Today it remains a region that is facing increasing cultural diversity with globalisation and the international migration of labour. The essays in this book therefore address issues that are at the heart of the development dilemmas faced by societies in the region. The questions that are posed and answered concern the choices that Southeast Asian societies must make now and in the future as they face the supposedly culturally homogenising forces of globalisation as well as the impact of rapid social and cultural change that economic growth has brought about in the region.

Chapter 1

Diverging Identities in a Dynamic Region

Brian J. Shaw

Introduction

Writing on the modern history of the region, Nicholas Tarling begins with the statement that 'Southeast Asia is marked by ethnic diversity' (Tarling 2001, 3). This statement recognises the importance of Southeast Asia as a cultural crossroads, a quality that has given rise to high levels of ethnic pluralism, not only between countries, sub-regions and urban areas but also at the local levels of community and neighbourhood. The foundations for such diversity can be traced back to the earliest migrations of early *Homo* populations, which settled in the region some 1.5 million years ago, characterised today as 'Java Man' by virtue of extensive fossil finds at Sangiran, in present-day central Java. However, notwithstanding the region's claim to importance in human prehistory during the Pleistocene Epoch, it is the more recent migrations occurring during the present Holocene period, specifically between 12,000 and 5,000 years BP, which are now credited by archaeologists as laying the foundations of the region's current ethnolinguistic diversity (Bellwood and Glover 2004). By that time, the area now occupied by present day China was an 'ethnic mosaic' with no less than five language families, namely the Sino-Tibetan, Austroasiatic, Tai, Hmong-Mien and Austronesian, making up the earliest populations of agricultural villages based on the cultivation of foxtail and broomcorn millet in the north and rice in the south (Bellwood 2004; Bellwood and Glover 2004).

Subsequent migrations, through Vietnam and the Malay Peninsula and via Taiwan and the Philippines, expanded these populations throughout the region, and beyond. Most spectacularly, the Austronesian dispersals that occurred between 5,500 and 1,000 years BP took such peoples into Melanesia, Micronesia and Polynesia, and ultimately as far as Hawai'i and Easter Island (Bellwood 2004). It was however in the fertile flood plains of the Southeast Asian mainland that the great agrarian kingdoms developed, based on intensive wet-rice cultivation systems that were finely attuned to the cycle of the prevailing monsoon (Wolters 1999). Here the highest caloric output per land area was achieved for cultivated grain, sustaining the economy and culture of successive agrarian empires that fostered the development of urban centres with their military power, religious institutions and artistic and cultural elites. However, as Owen (2005, 9) points out,

... these kingdoms rarely managed to establish long-term political, economic, religious or linguistic control over the uplands that surrounded them ... hill peoples, often ethnically and linguistically different from those below ... would seek protection from the next adjoining kingdom, manipulating tribute relationships to try to sustain their security.

Scholars have characterised such territorial arrangements as akin to the concept of the *mandala*, a Sanskrit term, which used in this way symbolises the waxing and waning of territories and group allegiances in the absence of firm boundaries and declared identities (Higham 1989). In a region where land was plentiful and population density still relatively low, rulers were more interested in the number of potential slaves that might be captured by a conquering army, rather than in the control of land *per se* (Jerndal and Rigg 1998).

Inevitably the history of the region has revolved around the stories of these 'kingdoms and super-kingdoms' such as the Mon-Khmer kingdom of Funan established at least two thousand years ago, the Khmer civilisation at Angkor, Champa in present day Vietnam, Pagan in Burma, Ayudhaya in today's southern Thailand, and the more recently documented sea-borne empire of Srivijaya (Tarling 2001, 10–15) (see Figure 1.1). Such predisposition has tended to downplay the fortunes of people living in highland areas, those who for the most part lived without written records. Moreover, lowlander prejudice towards these highland groups has defined their interrelationship in a classic 'hill-valley' dualism. Geography and ethnicity combined to produce minority groups in places such as present day Cambodia, Thailand and Vietnam, while in Burma the Shans, Karens and other minorities belied the concept of the nation-state. Yet, Milton Osborne (2000, 53) makes the point that hill peoples, while outsiders, 'played an important if highly varied role throughout the region. They could supply, or be a source of slaves, trade in forest products, or offer special skills such as the training of elephants.' However, while the highland ethnic minorities may have enhanced the glory of kings this most probably was not in conditions of their own choosing. As the Chinese emissary to Angkor, Zhou Daguan, saw fit to observe in the late thirteenth century,

Wild men from the hills can be bught (sic) to serve as slaves. Families of wealth may own more than one hundred; those of lesser means content themselves with ten or twenty; only the very poor have none (Freeman and Jacques 2006, 37).

The extent to which the emergence of the 'god-king' (*deva-raja*) endowed with mystical power and exalted status derives from the transfer of Indian culture and religion has been the subject of intense debate. Certainly the establishment of both overland and maritime trade connections between the sub-continent and the lands of 'Further India' immediately to the east fostered acculturation, but the prevailing wisdom now favours a process of 'localisation' whereby Southeast Asian societies adapted elements of both Indic and Sinic culture to meet their

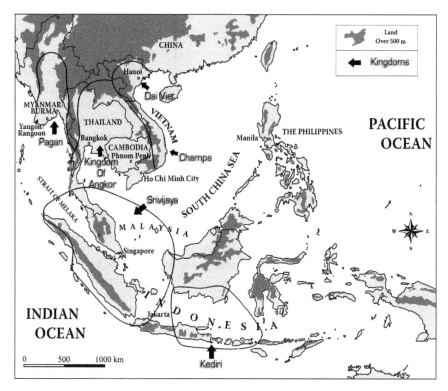

Figure 1.1   Southeast Asia

Source: Map by Bernard Shaw.

own needs (Hill 2002; Bellina and Glover 2004). Osborne (2000, 5–6) makes the point that the countries of Southeast Asia were neither 'little Indias' nor 'little Chinas', arguing the case for broad similarities across a wide area, through the adoption of the nuclear or individual family and the existence of linguistic unity particularly enhanced through the wide usage of Tai and Indonesian/Malay languages. But Osborne (2000, 8) then argues against his own thesis stressing 'the profound differences that do exist from place to place and between one ethnic group and another'. This apparent *volte-face* underlines the fundamental impasse that pervades the contents of this volume; to what extent should we celebrate the continuities that have formed this region's separate identity, or alternatively, stress the fragmentary nature of group and national identity and the challenges these pose for longer-term economic, political and social sustainability?

Maritime Incursions

Mention has been made of the maritime empire Srivijaya, which flourished between the fourth and thirteenth centuries, with a 'golden age' between the seventh and eleventh centuries. Unlike the mainland empires this was a thalassocracy centred in the area of present day Palembang, controlling the Melaka and Sunda Straits in both intra- and inter-regional flows of people, goods and services through port cities such as Aceh, Makassar and Patani. The ports functioned not only as commercial entrepôts, but also developed as political and cultural centres, described as 'port-polity' (Kathirithamby-Wells and Villiers 1990). Accordingly, this maritime empire displayed a high degree of ethno-linguistic variety with no dominant group, unlike the mainland empires with their 'insider-outsider' structure. This cosmopolitanism was enhanced by the nature of the monsoon environment, for in the days of sailing ships, suitable havens were particularly important in the Indian Ocean, South China Sea, Java Sea and Malacca Strait, when strong winds confined vessels to harbour until the change of season. During the waiting periods for the changing winds traders would stay for extended periods of time, building fortified camps near to pre-existing indigenous villages, and thus foster the development of hybrid settlements. Johannes Widodo (2004, 3) summarises these exchanges,

> The trading ships and immigrant boats were not only carrying people and goods, but also conveying cosmological and geometrical memories ... the different layers from different cultures have been super-imposed, adapted, and undergone process of indigenisation ... forming a truly blended cosmopolitan urban morphology and culture.

Srivijaya's success owed much to its tributary relationship with China, a country that jealously guarded its interests in the southern seas, which the Chinese termed *Nanyang*. During the early fifteenth century, the Ming Dynasty extended Chinese influence through the seven great expeditions of the Chinese-Muslim missionary navigator Admiral Zheng He, which further stimulated the development of cosmopolitan trading ports throughout the region. Anthony Reid (1993, 38) argues that the 'creative melding' of Chinese and Javanese marine technology in the wake of these expeditions prompted the subsequent expansion of Javanese shipping during the fifteenth century. A further beneficiary of greater Chinese presence in the region was the emergent port of Melaka, which became Southeast Asia's last great pre-colonial entrepôt. Thereafter, with the Portuguese conquest of Melaka in 1511, to be followed by Spanish, Dutch, French, British and American incursions, port cities were increasingly captured as footholds for Western powers, operating as 'beachheads of an exogenous system ... peripheral but nevertheless revolutionary' (Rhoads Murphey quoted in Reid 1989, 54). The subsequent 'colonial period' of Southeast Asia has thus been characterised as a 'watershed' in the region's development (Osborne 2000, 35–7), but in the context of maritime Southeast Asia it may be seen to have merely accelerated a process of ethnic, cultural and religious pluralism that had been taking place over previous centuries, most notably with the diffusion of Islam from the beginning of the thirteenth century.

While the mapping of European conquest in Southeast Asia ultimately revealed swathes of colour-coded territory depicting the possessions of the major colonial powers, impacts before the nineteenth century have been characterised as 'pin-prick imperialism' (Reed 2000, 69). Territorial possessions were largely confined to the coastal areas of insular Southeast Asia where a network of garrisoned ports supported local 'factories' engaged in overseas trade, but as Murphey (1989, 234) asserts 'Westerners and even their ships and trade remained less important than Asians and the trade they carried'. Traditional rulers continued to enjoy power, many of them willing partners in the 'divide and rule' policy' most successfully employed by the British. Numerically the number of Europeans was relatively small and the shortage of 'white women' in the colonies perpetuated that scarcity. Intermarriage, advocated by the Portuguese as a solution to this problem, produced communities of westernised Eurasians but this presence was essentially confined to coastal cities such as Melaka where their descendants survive to the present day (Hoyt 1993). The exception to this pattern was found in the Philippines where the Spanish conquistadors were driven, as in the Americas, by missionary zeal and the desire to secure colonised lands for the Spanish Crown. The Spanish replaced pre-existing indigenous beliefs and social systems with Roman Catholicism and the *encomienda* whereby conquistadors were granted a feudal form of trusteeship over the indigenous population (Ulack 2000).

In consequence, rather than being eclipsed by the arrival of a succession of colonial powers, a variety of other ethnic groups participated in the expanding trade networks. Shipping was conducted by Chinese, Javanese, Malays, Indians, Arabs, and increasingly as the eighteenth century drew on, by Buginese traders based in Makasar. The Chinese in particular prospered under the patronage of the European powers, not only as entrepreneurial traders but also as agriculturalists, craftsmen and miners, to the extent that Southeast Asia, in the eighteenth century, could be characterised as 'a zone of offshore production for China' (Carl Trocki quoted in Owen 2005, 27). Unfortunately, such propinquity had its downside when the Chinese, despite treaties assuring their protection, were caught between the interests of colonial powers, local rulers, rival elites and the indigenous populations. Most infamously in Manila 1603, and in Batavia (Jakarta) 1740, thousands of Chinese were massacred when colonial authorities overreacted to local tensions. Over the succeeding centuries eruptions of anti-Sinic sentiment throughout the region led variously to the imposition of immigration restrictions, sundry deportations and occasionally local pogroms against ethnic Chinese (Pan 1998). These atavistic tensions were most recently manifest in the 1998 Indonesian riots which scapegoated the non-*pribumi* (non-indigenous) Chinese population in the wake of the Asian financial crisis.

Good Fences Make Good Neighbours

Prior to the nineteenth century the main preoccupation of the European colonisers lay in the extraction of concessions and maintenance of trade routes within insular Southeast Asia. This situation changed with the demand imperatives that accompanied growing industrialisation in the West and a number of significant innovations that transformed relations between 'core' and 'periphery'. The gradual arrival of both the steamship and the international telegraph, followed by the opening of the Suez Canal in 1869, facilitated more effective communication between metropolitan-based governments and their distant colonies (Osborne 2000; Reed 2000). Attention turned towards the land-based empires of Southeast Asia where the imperial ambitions of the British and French found a new theatre for their political manoeuvres. The expansion of British power into northeastern India and the desire to secure imperial borders in the areas of Manipur and Assam challenged the Burmese Konbaung dynasty. In Indochina the French invasion of Vietnam as a precursor to opening up trade with China ended that country's independence. Squeezed between these imperial pincers, the kingdom of Siam, under the reigns of Mongkut and his son Chulalongkorn (Ramas IV and V), carried out a series of modernisations and made substantial territorial concessions in order to preserve the existence of the Thai state (Shaw 2001).

A casualty of these now rapidly colliding empires was the *mandala*, or notion of alliances based upon traditional yet shifting allegiances. The colonial powers demanded the security of frontier zones and the imposition of formalised boundaries, concepts that were alien to the indigenous rulers. The Treaty of London 1824 marked the division between British possessions in the Malay Peninsula and Dutch interests in Sumatra, thus splitting the area once part of the Srivijaya kingdom. Siam, which had made substantial trading concessions to Britain under the terms of the Bowring Treaty 1855, was subsequently forced to relinquish to the French its suzerainty of territories east of the Mekong, following the Pak Nam incident in 1893. Then, in 1909, Siam ceded to the British its southern dependencies of Kedah, Perlis, Kelantan and Terengganu. The British conquest of Burma, completed by the end of the nineteenth century, left former areas of Burmese influence beyond the border with India (Osborne 2000; Owen 2005). These and other boundary changes forged six political entities out of the kaleidoscope of pre-colonial states, namely British Burma; quasi-independent Siam; French Indochina; Spanish Philippines, British Malaya and the Dutch East Indies. While some subdivisions have taken place in the context of postcolonial nationalism, these political entities have for the most part survived to the present day (Elson 2004).

From an indigenous perspective these new boundaries were arbitrary and restrictive. At the stroke of a pen ethnic groups were relegated to the status of 'minorities' in their own traditional lands. Moreover, colonial authorities moved to 'regulate, constrict, count, standardise and hierarchically subordinate' the areas and peoples of the region (Benedict Anderson quoted in Owen 2005, 78). Psychological fences were erected between ethnic groups as the colonial practice of racial ascription classified them according to their economic or political potential within the colonial system. Some tribes were considered 'lazy, independent and turbulent' others 'low down in the scale of humanity' following the anthropological assumptions of the time. In Borneo the Iban were identified as good fighters and used to quell disturbances among other ethnic groups; the Chinese were encouraged for their trading skills; the Malays courted as political allies. Where indigenous populations were not considered suitable for specific economic activity, others were imported: Indians into Malaya's new rubber plantations, Tamils into tea plantations in Ceylon. The end results of such policies, legacies of colonial rule, were an identification of ethnicity with economic or political function, the forced, assisted or encouraged migration of ethnic groups and the displacement of indigenous peoples. The economic transformation of the region, undertaken as a colonial imperative, initiated a corresponding social and cultural transformation, elements of which are still enduring (Osborne 2000; Owen 2005; Reed 2000).

Yet, the most significant change that accompanied this intensification of colonial rule was a population explosion that began around the beginning of the nineteenth century, and indeed still continues to the present day. An early observer, John Crawfurd, wrote that Siam was 'inhabited by monkeys rather than people' (quoted in Hill 2002, 23), citing a population of 2.7 million for the whole country which also included territory in Laos. He gave estimates of some 5 million for Indochina, 6 million for Java, with 11 million for Indonesia as a whole, estimating the region's population at some 25 million in 1830 (Fisher 1964). These broad estimates were largely confirmed by other contemporary observers who noted the substantial population increases as the century progressed (Fisher 1964, 174–5). By the end of the nineteenth century, censuses conducted by colonial authorities recorded a regional total of over 80 million, and this figure more than doubled again by 1950 (Buchanan 1967; Fisher 1964; Hill 2002; Owen 2005). The reasons for this sudden increase after centuries of constancy are still open to debate, but change was not due to any major influx of population from outside the region with only some seven per cent of total regional growth attributed to such migration during the nineteenth century (Owen 2005, 224).

Scholars are sceptical about the suggestion that colonialism 'made life better' for the indigenous inhabitants of the region, although this sentiment undoubtedly found favour with contemporary colonial attitudes. Early population increases predated the arrival of modern medicines and general sanitation, and the transition from subsistence farming to cash crops may have even reduced food supplies in some places and periods. Fisher, writing in the early 1960s, argues the case for a suppression of internecine warfare under the law and order regime imposed by imperial powers. To back this statement he cites the incipient population growth experienced in the earlier subdued areas of Java and the Philippines, where ten-fold increases in population took place between 1829–1954, a trend then followed by more remoter regions of Burma, Indonesia and Malaya, as the century progressed (Fisher 1964, 172–3). Owen (2005, 227) broadly concurs with this analysis and suggests also that the incidence of famine may have fallen as food supplies could reach the hungriest areas through improved transportation. On balance, any benefits arising from internal pacification and famine relief should have impacted most positively upon smaller ethnic minority groups in the more remote parts of the region. Relativities notwithstanding, of all its impacts, the association between advanced colonialism and the onset of the demographic transition stands as the most durable. By the end of the twentieth century postcolonial continuation of this process had resulted in a regional population of over 500 million, with forward estimates suggesting a further 300 million to be added by 2050 (Hugo 2003).

Emergent Nationalisms

Prior to World War Two, the chances of successful nationalisms emerging within the artificially created boundaries of colonialism seemed a somewhat remote prospect. The region's elites had, by and large, identified with their colonial masters and a new 'middle class' of civil servants had begun to prosper in ways that had largely been denied to them under previous orders (Owen 2005, 212). The western concept of a nation state possessing unifying characteristics of culture, values and aspirations was at odds with these colonial creations. Robert Elson (2004, 27) makes the point that 'the host of minorities now corralled within colonial boundaries saw little benefit from advantaging the interests of majority national groups, and much to fear from their ascendancy'. Somewhat ironically, when fervent nationalism did reveal itself it was among those sections of the overseas Chinese community inspired by the anti-imperialism of Sun Yat-sen and politicised by the ensuing struggle between the Guomindang and the Chinese Communist Party. Elements of Southeast Asia's overseas Chinese population became fervent supporters and financial backers of mainland Chinese nationalism (Pan 1998), initiatives that did nothing to assuage the endemic apprehension and hostility held against them by other ethnic groups.

If the Japanese invasion and subsequent occupation of Southeast Asia between 1941–1945 was the historical turning point that shattered the myth of white supremacy, the midwives of nationalism took a variety of forms during the war and in the immediate aftermath of hostilities. Anti-Japanese resistance in Malaya had been largely organised by the Chinese through an offshoot of the Malayan Communist Party (MCP), an organisation that had links in Indonesia, and in Vietnam where the communists were mobilising against the French. The Burma Independence Army (BIA) had allied itself with the both the Japanese and returning British, fighting also against guerrilla forces organised by ethnic minorities. Within a few months of Burma's formal independence civil war broke out between former communist and non-communist allies. In the Philippines, communist-inspired resistance to Japanese occupation had been organised by indigenous groups as the Huk movement, and this group continued to fight on until well after the war's end (Berger 2004; Owen 2005). Generally, the regional complexities and local particularities of the immediate post-war era, set in the context of a developing

Cold War, overlaid the myriad class, ethnic and gender disparities apparent within Southeast Asian populations. Given the multiple confusions of the situation at that time, and subsequent crises and challenges, it is somewhat remarkable that the political template established by the colonial powers has proved remarkably resilient right up to the present day.

Writing in the early 1960s Charles Fisher (1962) posed the question 'Southeast Asia: The Balkans of The Orient?' His query was based upon physical comparisons in terms of common geographical fragmentation into peninsulas and islands; geopolitical similarities as both are located between greater and competing entities; and the 'extreme and baffling complexity of the ethnic and linguistic geography of both regions' (1962, 348). Fisher characterised both as 'areas of transition and instability – cultural and political fault zones' (1962, 347). His analysis of the region's emergent states identified 'three different kinds of component peoples', namely one or more advanced indigenous (majority) group; secondly, one or more traditionally less advanced indigenous group of mostly hill peoples; and, thirdly, one or more non-indigenous group such as the Chinese and Indians (1962, 360–61). While today's analysis would most probably discard the notion of relative advancement, Fisher's observation that 'in no case does a single language, or one particular religion, at present provide a common denominator' (1962, 361) still holds currency. Despite these pronouncements and the ongoing political upheavals of that time Fisher was ultimately optimistic about the future stability of the region, and we should take note that to date his confidence has been largely justified.

In the years immediately following decolonisation, the multi-layered characters of Southeast Asia's port cities as the particular repositories of the aforementioned third group, sat uncomfortably alongside the self-conscious building of national identity. For example, early Indonesian nationalism under Sukarno disavowed both Dutch colonial heritage and ethnic Chinese identity, evident in the official discouragement of Chinese signage on buildings, and suppression of the Chinese language (Pan 1998). In Melaka, the influences of Portuguese, Dutch and British colonisers and even the continued presence of local-born Chinese, were downplayed as the old port took central place in the creation of a distinctly Malay national identity that favoured the indigenous *Bumiputera* or 'sons of the soil' (Worden 2001). After separation from Malaysia in 1965, the ethnic Chinese majority city-state of Singapore was reduced to 'a head without a body', and responded to the challenges of ethnic pluralism by adopting a reductionist approach to the question of 'race'. The CMIO model (Chinese, Malays, Indians and Others), replaced the numerous dialect, religious and caste groupings from diverse geographical origins, while promising each collapsed entity a distinct and equal place within an overarching 'Singapore' national identity (Ooi and Shaw 2004).

As a consequence, the continuing stability of culturally pluralistic societies in Southeast Asia has not been achieved by the development of equally pluralistic governments. As Ooi (2006, 299) has elucidated,

Ethnic differences have been managed variously in Southeast Asian societies even as governments have pushed different development agendas. While such differences and varied policies have provided by and large, political stability in the region until the onset of the economic crisis in 1997, clearly such stability was based more on economic growth and wealth creation rather than inter-ethnic understanding and appreciation of ethnic diversity in the societies.

The modernisation paradigm of the 1950s and 1960s placed ethnic and social considerations in thrall of economic advancement. New mottos such as newly independent Indonesia's 'unity in diversity' (*Bhinneka Tunggal Ika*) characterised sentiment rather than reality. A competing influence, during the early phases of postcolonial independence, was the alternative communist paradigm whereby culture, ethnicity and even nationalism were seen as subservient to a higher, more egalitarian social order as revealed through dialectical materialism. Indeed, Keith Buchanan writing in 1967, foresaw the 'emergence of a wide spectrum of left-wing regimes' that while coloured by specific social and cultural characteristics, would have unity in the 'common fashion in which they confront the common problems posed by poverty and rapidly expanding populations' (1967, 160).

Constructing Identities

While Buchanan's prediction may now appear somewhat quixotic, the newly independent states of Southeast Asia, regardless of their political persuasions, have been distinguished by their attempts to forge national identity from above, following the path outlined by political theorists such as Benedict Anderson (1983) and Ernest Gellner (1987). The inscription of national values was conceived as a legitimating discourse in the face of widespread cultural, ethnic and social diversity. In some cases, notably Thailand, the retention and promotion of a hereditary monarchy, albeit constitutional rather than absolute, has provided a rallying point and most recently a 'network monarchy' of power structures linked to the palace (McCargo 2005). The extent to which this symbolic unity is a function of the personal veneration afforded to the world's currently longest serving head of state, Bhumibol Adulyadej (Rama IX), remains to be seen. Constitutional monarchy is found also in Cambodia, where its continued existence owes much to another long-term survivor, Norodom Sihanouk, whose latest incarnation is that of HM King-Father of Cambodia. Federated Malaysia also retains a constitutional monarchy but without the same degree of personification due to the five-year revolving system for the nomination of the Yang di-Pertuan Agong [Paramount Ruler] drawn from the nine regent sultans. Accordingly, within Malaysia, successful prime ministers have more frequently captured popular imagination. Unique within the region is the small oil-rich Sultanate of Brunei where Hassanal Bolkiah has ruled since 1967, combining the roles of head of state and head of government and projecting the notion of the Malay Muslim Monarchy (Melayu Islam Beraja) (Gunn 1997).

Where monarchs were absent, or had been eliminated as in Laos (Kremmer 1997), others emerged to take their place as figureheads for the nation. Indonesia, the Philippines and Singapore, although classified as republics, were subject to strong personalities as charismatic leaders invoked traditional values and strong moral codes in their exercise of power (Cartier 2000). In Vietnam, the memory of 'Uncle' Ho Chi Minh, architect of Vietnamese independence, continues to symbolise the unity of that nation almost 40 years after his death, and his embalmed body still lies in a mausoleum akin to that of Lenin's in Moscow. The creation of the Lao People's Democratic Republic (LPDR) in 1975 marked the beginning of a new era with the abolition of the 600 year-old monarchy, but subsequent socialist leadership has moved to accommodate traditional beliefs in a country where the Buddhist stupa Pha That Luang serves as the symbol of the nation (Stuart-Fox 1996). In Burma, the military government of General Ne Win repressed its own citizens and took the country into self-imposed isolation before being succeeded in 1988 by the brutal State Law and Order Restoration Council (SLORC, now State Peace and Development Council SPDC), which adopted the name Myanmar. Within this moral vacuum of successive Burmese administrations, the mantle of charismatic leadership has fallen on 'the voice of hope' 1991 Nobel Peace Prize winner Aung San Suu Kyi (1997), the elected PM who has largely been under house arrest since 1989.

Inevitably, the assertive personalities involved in the incipient process of nation building have been reliant upon the support of dominant ethnic groups. Indonesia's three hundred or so distinctive ethno-linguistic groups, spread over thousands of islands, were subject to the unifying language of Bahasa Indonesia [Indonesia Language] demographic engineering through the state transmigration policy and, in the eyes of minorities, excessive 'Javanization' of the government and its institutions. In Malaysia, following the 'racial riots' of 1969, the government instigated the New Economic Policy designed to promote the status of the majority Malay bumiputra population. In Burma, minority groups have actively resisted the dominant Burman majority since independence, in an ongoing struggle that has inhibited the development of that state. In the Philippines, Moro resistance against the majority Filipino culture still continues on the island of Mindanao. Within Indochina, ethnic minority 'hill tribes' collectively termed 'Lao Soung' in Laos, and in Vietnam highland groups such as the Muong, have been subject to marginalisation within these avowedly socialist states. These examples demonstrate the role of governments and state institutions as cultural and social gatekeepers, raising important questions about the politics of control in situations of ethnic pluralism. Moreover, all the above cases highlight the roles played by authoritarian regimes, or at best, 'pseudo- and low-quality democracies' (Case 2004, 76).

The question recently posed by Susan Henders, in the context of political regimes and ethnic identities in East and Southeast Asia, is 'Do Authoritarian Regimes Do It Better?' which leads to its corollary that 'the process of democratisation is itself destructive of interethnic accommodation' (2007, 3). In the same volume Daniel Bell makes the point that 'in a democratic system, political leaders must be sensitive to majority preferences and, thus, the language and culture of the majority group tend to be the only feasible basis of nation-building (Bell 2007, 31). Such a proposition will no doubt find favour with many of the region's 'low-quality democracies' characterised by less than free and fair competitive elections under universal franchise; freedom of speech, assembly and press; and where oppositions can criticise incumbents without fear of retaliation (op cit., 26). Indeed, David Wurfel (2007, 203) suggests that 'the Philippines is the only country in Southeast Asia having prolonged experience with democracy – that also has a multiplicity of ethnic groups'. A country of course unfortunately tainted by the period under the Marcos regime and still marred by ethno-religious conflict. By way of tentative summary, Henders suggests that the risk of ethnicised conflict, exclusion or hierarchy under democratisation is a function of the way ethnic identities were constituted under conditions of authoritarianism (2007, 10–11). Put simply, nations reap what they have previously sown.

Diverging Identities?

The notion of a uniform national culture developing in the top-down way envisioned by theorists such as Anderson and Gellner has been shown to be more than somewhat problematical in the Southeast Asian context. Moreover, such a hierarchical induction now represents an overtly modernist interpretation of events in the context of the more recent 'cultural turn' that has permeated the social sciences (Cook et al. 2000). The formerly fixed or 'natural' identity of 'race', so confidently prescribed by colonial authorities and willingly taken up by postcolonial nation-builders, can now be seen as a socially constructed concept. At the same time, the persistence of race referencing in the context of state-led social engineering has the continued effect of defining national discourse through its inculcation into public attitudes, perhaps best exemplified by Singapore's continued use of the CMIO categories. As Lai writes (2004, 9), 'the interchangeability of "race" and "ethnicity" (has) permeated all levels of society and is now common among officials, policy makers, academics and the public alike'. Judged in the context of this volume, with its emphasis on 'Diverging Identities', the Singapore categorisation might be seen as an inflexible classification of an increasingly fluid reality.

Along with its questioning of supposed absolutes, the cultural turn has also challenged the notion of 'history from above' placing an emphasis on previously hidden or 'subaltern' voices in the postcolonial context. There is a recognition that state ideology has in the past drawn selectively upon the complex histories and identities of its peoples, championing some, while sidelining others, ostensibly in the interests of national unity. This form of 'internal colonialism' has been seen at its worst in Burma, where the ruling military junta have stressed the 'Non-disintegration of the Union' as the paramount national resolution regarding nation building around the identity of the majority Bamars or Burmans (Tin 2005). Within Indonesia, Rizal Sukma (2005, 3) distinguishes between 'horizontal' and 'vertical' ethnic conflicts, the latter being the more intractable as ethnic groups with separate identities seek to secede from the central state, as demonstrated in the resistance shown by the people of Aceh and West Papua. In the Philippines the Muslims of Mindanao and Sulu, situated as Wurfel indicates (2007, 205) on the 'geographical, cultural and thus political periphery of the state', have been in long-term conflict with majority Filipino culture.

In discussing the struggles for autonomy in the Philippines, Miriam Ferrer offers the view that ethnic mobilisation can no longer be explained as a pure expression of primordial ties, but increasingly attributed to 'changing socio-economic and political factors and to the post-colonial restructuring processes in the context of global capitalism' (2005, 110). More broadly, this argument gains traction when placed within the context of an increasingly globalising world that has widened many pre-existing disparities and thus exacerbated long-established rivalries. The forces of 'postcolonialisation' in their various manifestations such as nation-building, state-led development projects and the rapid transition to a market economy have brought about an acceleration of social change, and are creating new and imagined communities that are potentially disruptive and which may well threaten the longer-term sustainability of the region. Overall, we should recognise that over recent times there has been a global trend towards greater rather than lesser cultural diversity within the region.

The chapters that follow will illustrate the various dimensions of this ongoing process at a variety of scales. In Chapter 2, Rahil Ismail gives an updated account of the challenges faced by the Singapore Malay-Muslim community in the district of Geylang Serai, generally acknowledged as the centre of that community's rich religious and cultural heritage. These challenges relate to a major redevelopment programme involving the demolition of iconic symbols, the resettlement of inhabitants, the temporary relocation of amenities and the building of new housing. The author relates recent events that took place during the fasting month of *Ramadan* in 2006, whereby the traditional role of the old Geylang Serai market was contested by the substantially redeveloped Kampong Glam district. However, this community has not been co-opted within the development of tourism related imagery, neither has it displayed a sufficient level of resistance to challenge the more tourist-oriented Kampong Glam district.

The pressure of redevelopment is also addressed in Chapter 3 where Ooi Giok Ling and Brian J. Shaw present three case studies of tourism oriented development in the islands of Sentosa in Singapore, and Langkawi and Pulau Pinang in Malaysia. The authors maintain that erasure of collective memory is made possible through the lack of identity in the absence of locally rooted populations in the face of central planning efficiency, best exemplified by the Singapore example. Where a network of public interest groups and concerned individuals is active enough to put forward alternative proposals, in the manner of recent mobilisations in Penang, successful resistance to the destruction of heritage through the creeping commercialism imposed by the demands of the global tourism and recreation industry is still possible.

In Chapter 4, in a somewhat different note but also concerned with the forces of globalisation, Mark Baildon writes on 'Singapore's Educational Reform as Post-Developmental Governance'. He shows how the state uses education to promote and sustain economic development in a relentlessly changing and increasingly competitive global economy. However, while balancing stability and order, the state also needs to create the conditions for innovation, creativity and experimentation. Singapore's pragmatism in this regard calls for practices that resist closure and social rigidity, and thus a removal of the limits set on dissent and the operations of civil society in order to foster economic growth and prosperity in the new economy.

Chapter 5 on the 'Morphogenesis and Hybridity of Southeast Asian Coastal Cities', by Johannes Widodo, documents the various influences upon the Southeast Asian region from a multiplicity of sources including India, Arab, Persia, China, Europe, Japan. The author presents a picture of largely peaceful and harmonious absorption of various cultures through exchanges which occurred in the South China Sea, Java Sea, and Melaka Strait; and area which he characterises as the 'Mediterranean of Asia', lying between the two sub-continents of China and India and the Pacific and Indian oceans.

A different perspective is offered in Chapter 6 where Kevin Blackburn's study of 'Nation-Building, Identity, and War Commemoration Spaces in Malaysia and Singapore', examines the way in which different approaches to nation-building and national identity have created contrasting war deathscapes in commemoration of the Japanese Occupation during World War II. In Singapore, state sponsored nation building has created a war deathscape at the Civilian War Memorial, in which all Singaporeans regardless of 'racial' identity remember their collective suffering under the Japanese. In Malaysia, notwithstanding its similar wartime experience, the massacres of tens of thousands of Chinese civilians are not collectively mourned, but throughout the country dozens of geographically scattered memorials are privately tendered by Chinese clan members.

In Chapter 7, Ambar Widiastuti looks at the process of globalisation from a personal perspective, asking what it means 'Being a Javanese in a Changing Javanese City'. Through an examination of the cultural event of *Sekaten*, the most popular Javanese ceremony in Yogyakarta, she shows how the ceremony has experienced shifts in its significance and functions, changing its location, being privatised and commercialised. In a similar vein to the earlier account of Geylang Serai, she concludes that culture is not merely about exterior manifestations and declarative statements, but encompasses deeply held values and beliefs, which underpin the symbolic or overt actions.

Chapter 8 entitled 'Re-imagining Economic Development in a Post-Colonial World: Towards Laos 2020' by Michael Theno, places the focus on the Lao Peoples Democratic Republic and the ongoing and unresolved combat legacy of the ethnic Hmong, together with the situations of numerous other ethnic hill tribes, each with their own language, customs, and culture. This is a story of contemporary social engineering through which the relocation and repatriation of Lao ethnic groups has been deemed essential to the process of poverty reduction, and geographical engineering through the process of extensive dam construction in the cause of development. The price of development, should it be achieved, may well be the loss of these distinct hill tribe identities formed over countless generations. The author posits that this is not without historical precedent.

The final Chapter by Nancy Hudson-Rodd, entitled 'When was Burma? Military Rules Since 1962' explores the complexities of a nation that has remained under military rule for over four decades. During this time, the violence committed by successive military regimes on Burma's civilian populations has squandered the country's potential to the point where Burma is now the most impoverished nation in the region. Historically there never existed a unified nation state of Burma, and since independence in 1948 the country has been plagued with complex ethno-territorial conflicts as ethnic groups have sought autonomy. However, as the author points out, ethnicity is used as a screen for other political interests. This is an exploration of nation state, people, ethnicity, place, human rights, and belonging in contemporary Burma. It is also one of hope for the future.

References

Anderson, B. (1983), *Imagined Communities: Reflections on the Origins and Spread of Nationalism* (London: Verso).

Aung San Suu Kyi (with Alan Clements) (1997), *The Voice of Hope* (New York: Seven Stories Press).

Bell, D. (2007), 'Is Democracy the "Least Bad" System for Minority Groups?' in Henders, S.J. (ed.) *Democratization and Identity: Regimes and Ethnicity in East and Southeast Asia* (Lanham: Lexington).

Bellina, B. and Glover, I. (2004), 'The Archaeology of Early Contact with India and the Mediterranean World, from the Fourth Century BC to the Fourth Century AD' in Glover, I. and Bellwood, P. (eds), *Southeast Asia: From Prehistory to History* (Abingdon: RoutledgeCurzon).

Bellwood, P. (2004), 'The Origins and Dispersals of Agricultural Communities in Southeast Asia' in Glover, I. and Bellwood, P. (eds), *Southeast Asia: From Prehistory to History* (Abingdon: RoutledgeCurzon).

Bellwood, P. and Glover, I. (2004), 'Southeast Asia: Foundations for an Archaeological History' in Glover, I. and Bellwood, P. (eds), *Southeast Asia: From Prehistory to History* (Abingdon: RoutledgeCurzon).

Berger, M.T. (2004), 'Decolonizing Southeast Asia: Nationalism, Revolution and the Cold War' in Beeson, M. (ed.), *Contemporary Southeast Asia* (Basingstoke: Palgrave Macmillan).

Buchanan, K. (1967), *The Southeast Asian World* (London: Bell and Sons).

Cartier, C.L. (2000), 'The Role of the State' in Leinbach, T.R. and Ulack, R. (eds), *Southeast Asia: Diversity and Development* (New Jersey: Prentice Hall).

Case, W. (2004), 'Democracy in Southeast Asia: How to Get It and What Does It Matter' in Beeson, M. (ed.), *Contemporary Southeast Asia* (Basingstoke: Palgrave Macmillan).

Cook, I., Crouch, D., Naylor, S. and Ryan J.R. (eds) (2000), *Cultural Turns/ Geographical Turns: Perspectives on Cultural Geography* (Harlow: Pearson Education).

Elson, R. (2004), 'Reinventing a Region: Southeast Asia and the Colonial Experience' in Beeson, M. (ed.), *Contemporary Southeast Asia* (Basingstoke: Palgrave Macmillan).

Ferrer, M.C. (2005), 'The Moro and the Cordillera Conflicts in the Philippines and the Struggle for Autonomy' in Snitwongse, K. and Thompson, W.S. (eds), *Ethnic Conflicts in Southeast Asia* (Singapore: ISEAS).

Fisher, C.A. (1962), 'Southeast Asia: The Balkans of the Orient? A Study in Continuity and Change', *Geography* 47, 347–67.

Fisher, C.A. (1964), *Southeast Asia* (London: Methuen).

Freeman, M. and Jacques, C. (2006), *Ancient Angkor* (Bangkok: River Books).

Gellner, E. (1987), *Culture, Identity, and Politics* (Cambridge; New York: Cambridge University Press).

Gunn, G. (1997), *Language, Power and Ideology in Brunei Darussalam* (Athens, Ohio: Ohio University Press).

Henders, S.J. (ed.) (2007), *Democratization and Identity: Regimes and Ethnicity in East and Southeast Asia* (Lanham: Lexington).

Higham, C. (1989), *The Archaeology of Mainland Southeast Asia* (Cambridge: Cambridge University Press).

Hill, R. (2002), *Southeast Asia: People, Land and Economy* (Crows Nest NSW: Allen and Unwin).

Hoyt, S.H. (1993), *Old Malacca* (Oxford: Oxford University Press).

Hugo, G. (2003), 'Demographic Change and Implications' in Chia L.S. (ed.) *Southeast Asia Transformed: A Geography of Change* (Singapore: ISEAS).

Jerndal, R. and Rigg, J. (1998), 'Making space in Laos: Constructing a National Identity in a "forgotten" Country', *Political Geography* 17: 7, 809–831.

Kathirithamby-Wells, J. and Villiers, J. (eds) (1990), *The Southeast Asian Port and Polity: Rise and Demise* (Singapore, National University Press).

Kremmer, C. (1997), *Stalking the Elephant Kings: In Search of Laos* (Chiang Mai: Silkworm).

Lai, A.H. (ed.) (2004), *Beyond Rituals and Riots: Ethnic Pluralism and Social Cohesion in Singapore* (Singapore: Marshall Cavendish).

McCargo, D. (2005), 'Network monarchy and legitimacy crises in Thailand' *The Pacific Review* 18:4, 499–519.

Murphey, R. (1989), 'On the Evolution of the Port City' in Broeze, F. (ed.) *Brides of the Sea: Port Cities of Asia from the 16th to 20th Centuries* (Kensington: University of New South Wales Press).

Osborne, M. (2000), *Southeast Asia: An Introductory History*, 8th edn (St Leonards NSW: Allen and Unwin).

Ooi, G.L. (2006), 'Ethnicity and the City in Southeast Asia' in Wong, T.-C., Shaw, B.J. and Goh, K.-C. (eds), *Challenging Sustainability: Urban Development and Change in Southeast Asia* (Singapore: Marshall Cavendish).

Ooi, G.L. and Shaw, B.J. (2004), *Beyond the Port City: Development and Identity in C21st Singapore* (Singapore: Pearson Prentice Hall).

Owen, N.G. (ed.) (2005), *The Emergence of Modern Southeast Asia: A New History* (Honolulu: University of Hawai'i Press).

Pan, L. (ed.) (1998), *The Encyclopedia of the Chinese Overseas* (Singapore: Archipelago Press and Landmark Books).

Reed, R.R. (2000), 'Historical and Cultural Patterns' in Leinbach, T.R. and Ulack, R. (eds), *Southeast Asia: Diversity and Development* (New Jersey: Prentice Hall).

Reid, A. (1989), 'The Organisation of Production in the Pre-colonial Southeast Asian Port City' in Broeze, F. (ed.), *Brides of the Sea: Port Cities of Asia from the 16th to 20th Centuries* (Kensington: University of New South Wales Press).

Reid, A. (1993), *Southeast Asia in the Age of Commerce 1450–1680: Volume Two Expansion and Crisis* (New Haven: Yale University Press).

Shaw, B.J. (2001), 'Mongkut (1851–1868) and Chulalongkorn (1868–1910) Kings of Siam', *Geographers Biobibliographical Studies* vol. 21, 65–71 (London; Mansell).

Stuart-Fox, M. (1996) *Buddhist Kingdom Marxist State: The Making of Modern Laos* (Bangkok: White Lotus).

Sukma, R. (2005), 'Ethnic Conflict in Indonesia: Causes and the Quest for Solution' in Snitwongse, K. and Thompson, W.S. (eds), *Ethnic Conflicts in Southeast Asia* (Singapore: ISEAS).

Tarling, N. (2001), *Southeast Asia: A Modern History* (Melbourne, Oxford University Press).

Tin, Maung Maung Than (2005), 'Dreams and Nightmares: State Building and Ethnic Conflict in Myanmar (Burma)' in Snitwongse, K. and Thompson, W.S. (eds), *Ethnic Conflicts in Southeast Asia* (Singapore: ISEAS).

Ulack, R. (2000), 'The Philippines' in Leinbach, T.R. and Ulack, R. (ed.), *Southeast Asia: Diversity and Development* (New Jersey: Prentice Hall).

Widodo, J. (2004), *The Boat and the City* (Singapore: Marshall Cavendish).

Wolters, O.W. (1999), *History, Culture, and Region in Southeast Asia* (Ithaca NY: Cornell University).

Worden, N. (2001), 'Where it all began': The Representation of Malaysian Heritage in Melaka', *International Journal of Heritage Studies* 7:3, 199–218.

Wurfel, D. (2007), 'Democracy, Nationalism, and Ethnic Identity: The Philippines and East Timor Compared' in Henders, S.J. (ed.), *Democratization and Identity: Regimes and Ethnicity in East and Southeast Asia* (Lanham: Lexington).

Chapter 2

'Di waktu petang di Geylang Serai'
Geylang Serai: Maintaining Identity
in a Globalised World

Rahil Ismail

Di waktu petang di Geylang Serai
Terang benderang sungguhlah ramai
Tua dan muda, miskin atau kaya
Semuanya ada terdapat disana[1]

Di Geylang Serai

This popular Malay song, *Di waktu petang di Geylang Serai* [In the afternoon in Geylang Serai], has an infectious jaunty tune that was fairly representative of songs popular in the Malay Peninsula in the 1960s. However, for the minority community of Singapore Malay-Muslims, the tune and the lyrics evoked an indelible part of the community's collective memory of time, space and with it identity. The Geylang Serai flats at Jalan Pasar Baru with its local wet market were situated in the heart of the larger Geylang district which sits at the Southeast of Singapore island (see Figure 2.1). It is generally acknowledged as the premier area associated with the Singapore Malay-Muslim population. Within this space and in the surrounding areas grew the embedment and intertwining manifestations of the community's rich religious and cultural heritage.

Geylang Serai truly emerged in the 1960s. It continues to be an area in which a minority community could gain the kind of in-group affirmation, support and growth that nourishes a community whose cultural practices are not nationally 'mainstream' but almost always perceived as the 'other': denoting 'difference' and at times a problematic difference.[2]

---

1   Essentially, the Malay lyrics exuberantly describe an afternoon scene in Geylang Serai, Singapore with its diverse visitors and wide selection of goods. This popular song was a hit by the singer Ahmad C.B. in the 1960s.

2   For many years the Malay population was identified in the main press with the reprehensible term the 'Malay Problem'.

As with most parts of Singapore, the larger Geylang area has been through several changes since the first major colonially-directed ethnic-based population resettlement programme. Since then, it has been transformed by post-independence resettlement and redevelopment projects as Singapore has become increasingly urbanised. However, through these changes the core area of Geylang Serai has remained an emotionally powerful focus for the Singapore Malay-Muslim community not just for its claim of privileged space but for its ability to symbolise a minority community's resilience and determination to maintain and cultivate their own cultural identity within a multi-racial society in an increasingly homogenised, globalising world.

Within this multiracial context that has seen the community change from being part of the majority group to, after the separation from Malaysia in 1965, a minority group, the current redevelopment plans for Geylang Serai have brought to the surface a series of challenges for the community and a heightened sense of administrative pragmatism and adroitness. Geylang Serai is an invaluable case study in the analysis of space, memory and identity considering 'the rapidly transforming societies where legacies of the past are constantly under the threat of erasure, where collective memories are ever so often revised, and where spatial identities and place meanings are always created anew' (Chang 2005). This chapter examines the current challenges, implications and consequences of the current physical transformations of Geylang Serai on community and memory.

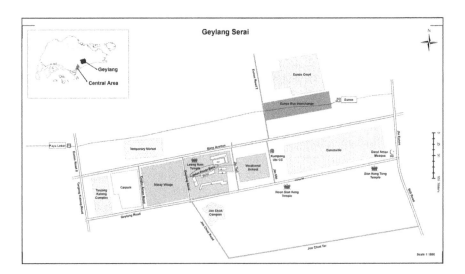

Figure 2.1   Geylang Serai

Source: Yan Peiyun and Kalpana d/o Manimohan.

In 2007, Singapore had approximately 3.6 million Singapore residents[3] of which the dominant group, at 76.8 per cent, was Chinese. Singaporeans, until the recent impact of globalisation and increased migration, have identified themselves within the nationally-known CMIO acronym: Chinese, Malay, Indian and 'Others'. Singapore Malays, as the largest minority group, are represented within the 'M' category though they can trace their roots to more diverse components such as Javanese, Boyanese, Acehnese and Bugis among many others (Mydin 2006).[4] British race management policies were not too concerned with the rich differences between these groups but deemed as Malays all those who speak the Malay language, follow Malay customs and practise the Islamic faith.[5] The latest census puts the Singapore Malay population at 13.9 per cent, with Indians at 7.9 per cent and 'Others' at 1.4 per cent of total Singapore residents.[6]

Maintaining Identity in a Globalised World

Just as Chinatown and Little India are denotatively ethnic-bound spaces, Geylang (together with Kampong Glam)[7] was partly a product of the British colonial divide-and-rule management approach that translated itself to a 'race-and-space' equation for its increasingly diverse subjects. Through British colonisation, the Japanese Occupation, Federation with Malaysia and later independence, Singapore's diverse communities have invested an incalculable emotional attachment and historical value to these spaces with their terminological elevation as ethnic enclaves or 'quarters' to (contested) heritage spaces marketed for interested tourists.[8]

As past and current debate has highlighted, the faces of these heritage spaces are negotiated through complex and myriad issues involving government departments, professional experts, interested members of the public and of course cold economic imperatives. However there are concerns such as the perceived cultural commodification, essentialisation, trivialisation and exotification of heritage sites and iconic buildings which can be construed as undermining Singapore's claim for

---

3   Based on the 2007 figure of 'Singapore Residents'. The total population for the same year stands at 4.59 million. See Singapore Government of Statistics at <http://www.singstat.gov.sg/stats/themes/people/hist/popn.html>.

4   According to the Singapore Department of Statistics, Malays are persons of Malay or Indonesian origin, such as Javanese, Boyanese, Bugis and so on. Later the term 'Malay' would include everyone from the Nusantara: the Malaya-Indonesian Archipelago.

5   Similarly, the various diverse groups from the South Asian continent and China were grouped together as 'Indians' and 'Chinese'.

6   See <http://www.singstat.gov.sg/stats/themes/people/hist/popn.html>.

7   'Kampung' means village and unless it is part of the official and historical designation, this paper will adopt the revised spelling of 'kampung', instead of the old spelling, 'kampong'.

8   See   <http://www.visitsingapore.com/publish/stbportal/en/home/what_to_see/ethnic_quarters/geyland_serai.html>.

a respectful multicultural environment for all (Ismail 2006; Yeoh and Teo 1996; Teo and Yeoh 1997). While this claim can be applied to some of the heritage sites in Singapore, such outcomes can have troubling amplification for a minority group's heritage spaces and identity given the limited, seasonal and marginal representation within the national landscape. These consequences are not bound to just the eradication or modification of bricks-and-mortar manifestations or iconic symbols but the resultant intangible, long-term sense of self and community. Essentially, the contestation over identity involves history, politics, economics, perceptions, assumptions and even un/informed perspectives, but almost always framed by a power dynamics where defining a space's future identity ultimately resides in the state and state-defined interests.

The challenges to Geylang Serai's architectural landscape and to the Malay Muslim community itself are presently indubitably tied to a major redevelopment programme that is to last to 2010 (A. Rahman 2005, 133). The plan involves the demolition of iconic symbols, the resettlement of inhabitants, the temporary relocation of amenities, the building of new housing and the skilled adaptation of activities to the current 'dislocation'. Without doubt, these upgrading exercises are timely but the implications of these changes are still surfacing with on-going debate on how they are impacting on not just a cultural identity but its spatial economic sustainability as well.

Crucially, however, the transformation process of Geylang Serai is insightful not just as an endogenous, sociological study of the spatial change of a minority community but simultaneously as a means of accessing more meaningfully the political and economic context within which the community has to function both as citizens and Muslims in the domestic and globalised world. As noted by Hussin Mutalib on the Singapore Muslims' quest for identity in a modern city-state, consideration of the intertwining historical, social, economic and political contexts are imperative as perhaps offering '… some insights about the relative impact on modernisation and globalisation upon ethno-religious, particularistic tendencies, and how modern multi-ethnic states today go about balancing competing minority ethno-religious pulls on one hand, and forging a common national identity and consciousness on the other' (Mutalib 2005).

In the current 'imagining' of the new Geylang Serai and its expected transformation, the dual local–global and state–community considerations have varying degrees or forms of articulation but they are nonetheless significant in the future of the community.

Geylang Serai: A Brief History

In Singapore's spatial comprehension, there is a difference between Geylang and Geylang Serai with the latter being closely identified with the Malay community while Geylang itself is a larger area covering diverse ethnic groups and economic activities.[9] Through the decades, Geylang Serai formed the heart of the Malay community or, as noted by the National Archives publication on the history of Geylang Serai, it was 'The Malay Emporium of Singapore' (National Archives 1986, foreword). Its reputation is well-established among Malay-Muslims and visitors from Malaysia, Indonesia and Brunei with the 1970s representing the heyday for tourism from these countries.

Fundamentally, Geylang Serai's reputation lies with it being one of the oldest Malay settlements in Singapore with early Malay and *Orang Laut* [Sea People] settlers setting-up home along the Geylang River. The movement began with the 1840s British removal policy as it began dispersing the 'Malay floating village at the mouth of the Singapore River because of its obstruction to port traffic' (National Archives 1986, 16), with settlers moving inland and into Geylang. Consequently, these settlements evolved into *Kampong Kelapa* [Coconut Village], a name that reflected the abundance of the cash crops and the village's related economic activities. By the mid-nineteenth century, 'Gaylang' village or *Pulo Gelang* [Geylang Island] began to appear on maps with the seeming disappearance of the island as the progressive reclamation of Kallang Basin started. A shift in economic activity came with investments from the influential Alsagoff family in the area through their vast Perseverance Estate which specialised in the cultivation of 'serai' or lemon grass. As a source of citronella oil and other cosmetic uses, 'serai' products had a ready Western market but the Citronella Press Factory terminated its business in the 1890s. The settlement continued through assorted agriculture-based economic activities that included poultry rearing, vegetable cultivation and, of course, coconut-related activities (National Archives 1986, 17–20). Amidst the changes, the 'serai' plantation had already made its mark on the area and Singapore's history through the bestowment of its name.

There are various contesting versions of the origins of the name Geylang Serai but most seem to settle on the following plausible explanations. 'Geylang' could possibly be an evolution (or corruption) of the Malay word *gelang*, which means bracelet, or the word *kilang*, which means a factory or a press-mill related to the extraction of oil from copra or citronella oil from 'serai' plants. 'Serai', without doubt, refers to the lemon grass plantation. The close identification of the area as Geylang Serai and with it the Malay-Muslim population continued through the next century, with sub-sets of the accompanying areas including, for example, *Kampong Ubi* [Tapioca Village].[10]

Throughout the Japanese Occupation and the chaotic immediate postwar years, the larger Geylang grew into a mélange of Malay and Chinese *kampungs* with an increased population and strained infrastructure. However, as noted by Warwick Neville in 1966, Geylang Serai had truly established its identity as a Malay space:

9　Though among the Malay community, it is generally understood that when one mentions or alludes to living or visiting Geylang, it means almost always Geylang Serai.

10　The latter's name emerged from the Japanese Occupation policy of reserving the district for vast tapioca plantations as a means of replacing rice with tapioca as a wartime imperative.

> The Malaysians [*sic*] are most heavily concentrated in the Geylang Serai district
> where the main nucleus of settlement occurs outside the administrative city limits.
> The Malaysian [*sic*] population is almost wholly lacking in Chinatown and only
> minor nuclei occur in the northern section of the Rochore area; a significant
> proportion of the Malaysian [*sic*] population is located in the Southern Islands
> (1966, 248).

Changes in migratory patterns also contributed to this as the Chinese began
moving out as more Malays moved into Geylang Serai. What should be noted is
that although Geylang had Malay settlements or that Geylang Serai was and still
is Malay-dominated, the Chinese presence was and is still visible and thriving.
There were ethnically diverse *kampungs* with Chinese and Malay families and
houses of worship co-existing just as there was the prominent Leong Nam
Temple in the heart of Geylang Serai until 2006, and currently other temples
along Changi Road.

1960s: Change and Consolidation and Change again ...

Indubitably, the 1960s was the decade in which Geylang Serai came into its own
and consolidated its reputation as a viable and thriving economic and communal
anchor for the disparate Malay *kampungs* in Geylang and around the island. The
decade did not start well as Geylang became part of the official narrative of the
1964 'race riots' which affected the area significantly.[11] Since then the official
interpretive lessons of the event have become a touchstone for every conceivable
national dialogue or action with gory pictures of the dead and injured in Geylang
as evidentiary props. There were real concerns of Geylang becoming a recurring
flashpoint after 1964 and especially during the difficult months preceding the
separation of 1965. A 1967 assessment noted:

> At the time of separation one of the principal fears of the Singapore Government
> was that there would be a renewal of communal violence especially in areas like
> Geylang Serai, the scene of such outbreaks in 1964 (Leifer 1967, 289).

Concurrent with this insight was the need for meaningful and substantial pre-
emptive actions that were both defensive and forward-looking in their conception,
execution and completion (Kong and Yeoh 1994).

This was the acknowledgement, from the government's point of view, that
within the plans for urban renewal and settlement, Geylang needed attention and
with it 'management'. There was a palpable concern that although the various
kampungs may be of differing standards, there was a lack of amenities and

---

11   The disturbances began on 21 July 1964, preceded in April by a bomb exploding
in Geylang Serai as part of the early tentative actions during Indonesia's Confrontation with
Malaysia from 1963 to 1966.

increasing population could sink the area into an 'urban slum' (National Archives 1986, 27). With the 1964 formulation of the urban renewal policy, and the 1966 establishment of the Urban Renewal Department of the Housing and Development Board (Kong and Yeoh 1994), the extensive, complex and integrated urban renewal programme of the 1960s consolidated Geylang Serai's status and fortunes. As noted by Neville again:

> Government policy has played an increasingly significant role in designating areas for settlement and resettlement particularly under large-scale schemes to provide high density housing. Most significant instance influencing the existing ethnic distribution was the reservation of land for Malaysian [*sic*] settlement in the Paya Lebar-Geylang Serai area (1966, 249–50).

These massive programmes and their consequent outcomes elevated the community's emotional and spatial investment into the core area of present-day Geylang Serai. The area assumed a modern physicality and permanence of 'bricks-and-mortar' that formalised its cultural and economic status while community-affirming activities continued to be embroidered in the fabric of the emotionally charged collective memory. It is this seemingly intangible but galvanistic memory that has become the most potent companion or assessment of the space's future and the Malay-Muslim sense of self. Concurrently, it has provided the biggest challenge to the future direction of re-imagining what the Malay-Muslim community ought or should be in a multiracial, supposedly secular society as bound by national (and economic) imperatives.

The iconic buildings of Geylang Serai, which most people came to recognise as the heart of the space, but which were demolished in 2006, originated in the 3 August 1963, $3.7 million project that planned for three blocks of flats, a new market (*pasar baru*) and other associated amenities.[12] The government had earlier bought a site of 400,000 square feet 'bounded by Changi Road, Jalan Alsagoff, Jalan Rebong and Jalan Turi' and together with the Housing Development Board (HDB) planned to 'convert the area into a modern housing estate with new shopping centre and recreational centre'. This was all part of the plan to 'make the kampong area of Geylang Serai part of modern Singapore' (National Archives 1986, 27). The opening ceremony took place in April 1964 and construction on the site began the same month. It was completed in 1965 with the new flats housing approximately 4,000 residents (A. Rahman 2005, 83). The plan was complemented by a massive and the highly-successful resettlement project of the 1960s and 1970s in the outlying areas as Geylang Serai itself was marked for both low-cost housing and light industries. The Jalan Eunos Estate (later Eunos Crescent Estate) was completed in 1976 atop what was once a large swathe of *kampungs* with their highly-knit and vibrant communities.

---

12 Two other blocks of flats were added later.

The plan was for a modern satellite town as the former *kampung* residents themselves were resettled into other modern satellite towns of HDB apartment flats with all the modern conveniences. The internal diaspora of the Malay population saw movements into areas such as Bedok, Marine Parade, Kallang and Aljunied.[13] The move raised concerns regarding the loss of strong communal ties and the *kampung* spirit of family and community, especially with the dislocation and diffusion of the Malay population within the larger ethnic group that made up multiracial Singapore. The sense of displacement was palpable, one of the outcomes being the heightened emotional investment in Geylang Serai with its new buildings and shopping and entertainment amenities catering to Malay-Muslim needs.[14] Thus Geylang Serai's role as focus and anchor of the Malay-Muslim community from around the island became further embedded in the completed site which housed the Geylang Serai Redevelopment Scheme.

Geylang Serai has evolved since the 1960s to include the construction of the popular Joo Chiat complex in 1983, the struggling $10 million Malay Village 'showcase' in 1989, the impressive Darul Aman Mosque (corner of Changi Road and Jalan Eunos) in 1986, the completion of the HDB Eunos Estate and the Housing and Urban Development Company (HUDC) Eunosville estate in the 1980s. By 1996, the formalisation of the *Ramadan Hari Raya Bazaar* throughout the Muslim fasting month of Ramadan underlined not just the formal/informal aspects of this emblematic and much-loved seasonal food and shopping phenomenon, it cemented Geylang Serai's status and reputation as the centre of Malay-Muslim identity and memory.

By 1986, in the history of Geylang Serai published in conjunction with a pictorial exhibition organised by the Singapore National Archives on Geylang Serai, the conclusion was

> ... Geylang Serai today has developed into a predominantly Malay shopping district located at the old Malay residential areas. It is well-known as a place attracting not only Singapore Malays but also visitors from Malaysia, Brunei and Indonesia ... It has also become a meeting place for all Malays living in different parts of Singapore (1986, foreword).

The exhibition was preceded by the announcement by the Urban Redevelopment Authority of Singapore (URA) in 1985 of a Master Plan for Geylang and by 1994, 'Geylang the Planning Report' noted under 'Strengths and Opportunities':

---

13   Bedok, for example, is noted for its large Malay population but it is not a Malay dominated area.

14   Apart from the political electoral diffusion and its consequent implications, the population dispersal also meant navigating an everyday reality that included wet markets that sell pork. The Geylang Serai market was and is still the only 'halal' wet market as it does not sell pork.

The Sub-Regional Centre, commercial areas along Geylang Road and Joo Chiat Road and the industrial estate at Kallang Way and Kampong Ubi serve as employment centres. There is potential for redevelopment of the vacant land and low-rise industrial areas around the Paya Lebar Sub-Regional Centre[15] (Urban Redevelopment Authority 1994).

Not unexpectedly, the continual and integrated urban and national economic planning did not mean Geylang Serai would be left untouched. Furthermore, by the 1980s the general wear and tear on the Geylang Serai core district also resulted in the area coming in serious need of an upgrade or a complete change. The famous Geylang Serai market, still affectionately referred to as *Pasar Baru* [new market], was decidedly 'old and run-down' (Speech by Mr Mohamad Maidin Packer Mohd. 2004). Crucially, when redevelopment plans were finally announced, the same sense of anxiety that accompanied the 1960s changes surfaced in the 1980s. This anxiety, however, was tempered with the knowledge that the Singapore government had a respectable track record in past renewal and upgrading projects with its methodical procedures, its collaborative methods and reasonable compensatory packages. It is fair to surmise that most inhabitants of Geylang Serai that were marked for either rehousing or temporary dislocation would agree with the community newspaper's claims that '[A] key priority of the Government and Geylang Serai community in this development has been to retain the unique ambience and the enterprising spirit of Geylang Serai' (*Kampung Ubi News*, October 2005, 2).

However, there are crucial but not necessarily visible questions lying under the surface of a generally positive response to the announcement. Essentially, what are the other extra-contextual, non-quantifiable outcomes that might emerge; what exactly is meant by 'unique ambience', or what aspects of this 'ambience' will be retained or changed beyond recognition; who decides what is authentic; will the retained 'traditional' trades and lifestyles validate and affirm identity respectfully or be touristified to death; and will the changes see the making of another disappointment such as the intended tourist attraction, the Malay Village? Fundamentally, will there be community or commodification? The hope of a long-time stallholder on redevelopment is that it 'may ensure that tradition continues within a modern set-up' (A. Rahman 2005, 35), but tradition has often fallen under the demands of political and economic pragmatism with a certain degree of elasticity as to what 'pragmatism' and even 'tradition' should entail.

---

15   The 1994 report can be found here: <http://www.ura.gov.sg/dgp_reports/geylang/main.html> and quotes accessible at <http://www.ura.gov.sg/dgp_reports/geylang/pa-strop.html>.

After 42 Years: A Timely Change

Any form of change has degrees of difficulty and consequence, and more so with a minority community: its sense of privileged space and the lifetime of cultivated and nurtured collective memory. The intangibility of memory is often closely wedded to physical tangibility and with Geylang Serai, the iconic buildings and architectural landscape are deeply significant reminders of personal, cultural and even political identity. Will this affirmed space lose not just its authenticity by the relocation of its flat-dwellers but also the source of life that has taken years to cultivate?

It is not incorrect to assert that the plans for the new-look Geylang Serai provoked much debate and speculation within the community with one of the early (incorrect) lamentations being the demise of the *Hari Raya Bazaar* during the relocation period. The plans which involved demolition of the now iconic Geylang Serai flats and with it presumably the end of the *Hari Raya* activities concentrated at the foot of flats, created a sense of deep impending loss together with stoic acceptance of redevelopment inevitability. The *Hari Raya Bazaar*, which had grown in volume and space over the years, to spread out from the car parks of the Geylang Serai flats to the area bound by Tanjong Katong Complex, had become an institution: that it would not make an annual appearance would be inconceivable.

Fundamentally, the change is a 'comprehensive development' (Speech by Mr Mohamad Maidin Packer Mohd. 2004) and an upgrading of existing facilities and not the conservation of an area specifically targeted for tourist consumption. The redevelopment plans are ostensibly for the improvement in the quality of life of its residents, business owners and those who have a stake in the renewal and continual survival of the district. Undoubtedly it was hoped that the expected changes and improvements to the area would attract more tourists or non-Malay visitors to the heartland of an ethnic 'quarter' and hopefully revive the fortunes of the struggling Malay Village (see Powell 1997). The Malay Village represents the 'in/congruence of official and populist interests, the remaking of identity, the invention of tradition, the commodification of culture' (Ismail 2006) and is a byword for caution, consideration and the perils of re-imagining an 'other'. This is especially so when that 'other' is an ethnic minority re-imagined through the romantic notions of what stands for the authenticity of the ethnic 'other' by the powers-that-be.[16] While there was and still is concern with the possible 'erasure of memory' (Powell 1997, 94) through the demolition of iconic buildings and their associated uses, the plan for Geylang Serai is a comprehensive and integrated revitalisation programme to rebuild new homes on the old space, with a new, modern market and the development of peripheral businesses and buildings to ensure a certain degree of continual economic viability. It is not a piecemeal or specifically tourist-directed project such as the Malay Village but one for the

---

16   The Malay village was constructed in the Malay district of Geylang Serai in 1989, costing $10 million. Developed by the Housing Development Board (HDB) it never fulfilled its potential to 'showcase' a Malay village and had been through several makeovers.

stakeholders of the community that also includes the businesses and interests of Chinese and Indian members of the community.

In a February 2003 news release by the National Environment Agency (NEA) issued jointly with the HDB, the plans for the 'Selective En Bloc Redevelopment Scheme [SERS][17] and Hawker Centre Upgrading Programme at Geylang Serai' were announced. The key components: the 38-year old Blocks 1 to 3 of Geylang Serai at Jalan Pasar Baru had been identified for SERS with the five buildings demolished and residents compensated and/or relocated to Eunos Crescent (later Eunos Court) which is within walking distance of their present homes. The 'old and dilapidated' market would be demolished and rebuilt under the Hawker Centre Upgrading Programme (HUP) with stallholders offered replacement stalls in the temporary market to be set-up within sight of the old market. The integrated demolition-rebuilding and temporary relocation of some businesses were explained with the promise of an improved neighbourhood with exciting opportunities for continual development and a proposed return date set for late 2008.[18]

> The demolition of the Blocks 1 to 5 Jalan Pasar Baru/Geylang Serai presents a good opportunity for the redevelopment of the existing Geylang Serai Market. A prominent site fronting Changi Road can be freed up for the rebuilding of the Geylang Serai Market with better and more spacious design, improved ventilation and fire safety features. This will also sustain the vibrancy and character of the area (NEA and HDB 2003).

What was comforting was the announcement that '[S]pace for festive activities, such as the Hari Raya bazaar, will be integrated into the redevelopment of the Geylang Serai estate' (NEA and HDB 2003). Overall, the proposed 'new look' will constitute a new two-storey market with improved ventilation, fire-safety measures, pedestrian walkway and overhead bridge link to Joo Chiat Complex. Four hundred and forty-seven flats will be ready by 2010 to coincide with the completion of other proposed amenities in the redeveloped area.

The HDB housing project to build the 447 flats named 'Sri Geylang Serai' was launched on October 2006 with over 1,800 applications submitted by November. Much was made of the convenience of these new flats, built on the old Geylang Serai plot, as being accessible to markets, trains and schools, which will undoubtedly

---

17　SERS 'was announced as part of the Government's renewal plan to rejuvenate and intensify developments in older or mature estates. Old blocks of HDB flats with redevelopment potential and which are deemed as under utilised by today's planning standards are identified, demolished and newer blocks are built in replacement.' See <http://www.housingauthority.gov.hk/hdw/ihc/pdf/ihc04hd.pdf>.

18　While stallholders of the market were offered places for business in the temporary market, the shopkeepers existing on the ground floor of the five blocks of flat were not given replacement units but compensated to find other premises for business on their own. See <http://app.nea.gov.sg/cms/htdocs/article.asp?pid=1948>.

increase their value when the integrated redevelopment programme is completed in 2010 (Hussaini 2006b). Not unexpectedly, the issue of sentiment played a part with Malay-Muslim applicants with one seeing her effort as a form of '*kembali tinggal di Geylang Serai*' ['returning to live in Geylang Serai'] (Mohd. 31 October 2006) 40 years after she was resettled from the area. Her sentiment was in the fine Malay tradition of *balik kampong*, returning to the village/home where one has familial ties and cultural roots. For Malay-Muslim applicants, the reality, however, is that with this new project, the ethnic quota which restricts the number of ethnic groups in HDB estates would apply and with it possible demographic challenges to the previous *status quo* (see Chih 2002; Shaw and Ismail 2006a).

What is constantly repeated too is that the new market building is to be constructed in 'an architectural style reminiscent of rustic Malay dwellings' (*Kampung Ubi News*, October 2005, 3): a seeming reassurance that there will be no 'erasure' of identity but instead an affirmation of the Malay-Muslim identity. Without doubt, the hope is for the distinct building to evolve into an iconic landmark, especially significant in the relentless face of globalisation. Iconic community structures can 'act as counterbalance to transnational iconic projects that often threaten to disassociate the local society. Because they are underpinned by social and civic relations of locals, community icons ensure greater cultural diversity, participation and the reproduction of life spaces' (Ho 2006).

Insightfully, the vein running through the reports of the proposed plans in the major news media was the reference to the subtle 'reminders' of Geylang Serai as denoting the 'other' within the Singapore CMIO context. Whether it was an unconscious act of infantilisation or the internal touristification of an ethnic minority, descriptions tended to cover the superficial three 'F's of 'food, fashion and festival' (Ismail 2007, forthcoming), aspects of multiracial understanding of the country of its citizens. Together with 'rustic', it was retaining the 'charms' of the market (Anon. 27 February 2006; A. Rahman 2005) and 'the whole area comes alive' especially during Ramadan (Kwan 2007). While acknowledging the heritage of the area with an architectural homage, the spectre of the Malay Village's fate with its perceived obsession in 'stylised' (National Archives 1986, foreword) or fake authenticity did prompt quiet musings.

For many Singapore Malay-Muslims, a visit to Geylang Serai during Ramadan is part of an annual ritual that consolidates, perpetuates and renews the sense of self and community while investing further in the collective memory of the Geylang Serai. Significantly, an event in which a minority community is most comfortable and confident of its identity has also drawn national interest in its tourist value. Like a similar and age-old Bussorah Street seasonal food bazaar in the Arab district of Kampong Glam (Ismail 2006), the Geylang Serai bazaar is gradually marketed in tourist brochures and websites as the best time to visit this particular 'ethnic quarter': part of the 'Uniquely Singapore' marketing image.[19] Without

---

19   Geylang Serai is conveniently described as an 'enclave' with negative associations of Malay-Muslims' apparent intransigence to integrate into that terminologically ambiguous

dismissing totally the value of acknowledging this event as having a significant tourist value, there are necessary questions on recognising a community's event as a form of watching the 'natives at play'. As a form of accessing or understanding meaningfully the culture of the 'other', this has all the dangers of 'touristification' and 'exotification' that obscure the real challenges and realities of being a minority community in a multiracial society. Underlining these reservations are apprehensions: will adapting and embracing this partly exogenous-directed change compromise attempts to protect, preserve and promote ethno-religious identity in apparently secular and increasingly globalised Singapore?

It might be worth noting that the 2 November 2006 commentary in the only Malay daily in Singapore, the *Berita Harian* [The Daily News], under the headline '*Citra Melayu pada Geylang Serai yang lebih indah*' ['Malay narratives for a more beautiful Geylang Serai'] (Anon. 2 Nov. 2006), articulated hope in its last paragraph for an aesthetically improved Geylang Serai that could even be an attraction for tourists to see for themselves the life and culture of the Malay community.

The Remaking of Geylang Serai

The remaking of Geylang Serai started in earnest in 2005 and proceeded more or less according to schedule (compare Figures 2.2 and 2.3). The major milestones and events were closely covered by the Malay media with the emphasis on both the loss of memories while looking forward to new beginnings and opportunities. Throughout this process and since the announcement of the plans in 2003, the questions for the community and interested members were articulated publicly, in closed-door sessions with government representatives or in familiar circles (Anon. 24 February 2003). There were of course grievances expressed such as the increased cost of rent and conservancy charges in the new temporary market (A. Rahman 2005, 147), the perceived inadequacy of the compensation package among some, the perceived lack of support for shopkeepers who were not given temporary or replacement shops in the new Geylang Serai (Anon. 21 February 2003), and the loss of close cross-ethnic friendships forged over the years. In the word of a resident relocated to the new Eunos Court, it was all very '*sedih*' ['sad'].[20] The Malay television channel marked the occasion with a Malay drama series 'Geylang Si Paku Geylang' revolving around the lives and loves of fictionalised residents during the last days of Geylang Serai.[21]

---

but expedient term 'mainstream' but condescendingly approving if it can be touristified as an 'ethnic quarter'. See the Singapore Tourist Board marketing of the area under the slogan 'Uniquely Singapore' at <http://www.visitsingapore.com/publish/stbportal/en/home/what_to_see/ethnic_quarters/geyland_serai.html>.

20   Conversations with Encik Hussain Saaid in 2006 and 2007.
21   The drama series broadcast in late 2006.

Figure 2.2   Hari Raya Bazaar, October 2005

However, amidst all this, the efficient implementation and adroit responses to unexpected outcomes that had marked the execution of the redevelopment programme so far. The last *Hari Raya Bazaar* of the old Geylang Serai took place in October 2005; the residents observed the last *Hari Raya* celebrations on 3 November 2005; the move to Eunos Court began in earnest in late 2005 as the old market had its last day on 26 February 2006 in the glare of television cameras and curious camera-toting visitors. The cranes moved into Geylang Serai in April 2006 and this was marked on the front page of the *Berita Harian* with the declaration '*Pasar Geylang Serai tinggal kenangan*' ['Geylang Serai market only a memory'] (Hussin 2006). The construction and completion of, and move to the new temporary Geylang Serai market went in tandem with the above actions and its opening was an event marked by a full-page invitation in the *Berita Harian* to members of the public to the 'Opening Ceremony' on 8 April 2006 (Anon. 8 April 2006).

Together with approving reports in the media by both stallholders and visitors, the market was toured by Members of Parliament and a former Prime Minister who paid a visit to a 'popular market' (Anon. 6 November 2006). Early hiccups such as the provision of a controlled pedestrian crossing, a taxi stand, extended parking spaces for visitors and stallholders and also the provision of a *surau* (small prayer hall) for Muslim stallholders and visitors were attended to swiftly (Hamzah 2006a). This was followed by the introduction of weekend *pasar lambak* ('flea

Figure 2.3   Demolition of Geylang Serai flats, September 2006

market'; Hashim 2006) and the surprisingly regular use of the open space next to the market for media events, popular concerts and other community events that included Chinese religious celebrations. Concomitant to this was the report exhorting stallholders (and visitors) to remain vigilant in keeping the market neat and clean as well as to refrain from employing undocumented workers on the premises (Hussaini 2006a).

Across the road, Joo Chiat complex, which had symbiotic economic relations with the market, especially during the weekends, was affected by reduced shopping traffic and hence had to close businesses early. This in turn attracted less-than-desirable elements such as substance abusers and sex workers from the adjoining Joo Chiat Road district (see Shaw and Ismail 2006b) moving into what was originally the forbidden space of Joo Chiat complex (Mohd. 17 May 2006). Reports in May 2006 listed concerted responses to this spatial encroachment with police raids on Joo Chiat Complex (Mohd. 21 May 2006). There was concern that these 'extra-curricular activities' should not assume permanence. Police action was complemented with proactive measures by grassroots leaders and merchant associations in discussion with the Member of Parliament to '*kecohkan*' (create a sense of excitement; Saparin 2006) Joo Chiat Complex through the organisation of special shopping events and promotions. Essentially, the Complex needed to maintain a viable economic holding pattern until the opening of the new market in

2008 and the completion of the replacement flats, the 'Sri Geylang Serai', in 2010. Busy shopping weekends at Joo Chiat Complex might attest to some degree of success of giving a *'wajah baru'* ('new face'; Hambari 2006) to the complex.

The Challenges to Geylang Serai

The contestation issues discussed above on the politics of space and the remaking of an ethnic minority identity produced a subsequent test case of sorts. This was over Geylang Serai's most high-profile event: the focus of celebration of the holy fasting month of Ramadan. Ramadan in 2006 was to be the first without the old Geylang Serai market, and in the presumed vacuum the progressively conserved but ritzy Kampong Glam district saw an opportunity to fill an expected need in providing an alternative space to celebrate the event. An expanded Ramadan role was envisaged for Kampong Glam beyond the 'seasonal space' of buying and selling food to a full-scale affair to either replicate or replace Geylang Serai (Ismail 2006).

Indubitably, observers were curious to witness the head-to-head outcome of these two major Malay-Muslim spaces as Geylang Serai resurrected the *Hari Raya Bazaar* at its temporary premises and was bigger than ever. The *Hari Raya* lights went up as before, the seasonal stallholders came with their wares and so did the crowd as early apprehensions of its economic viability evaporated swiftly. As described, it was a: '... giant food and shopping paradise that runs from Paya Lebar MRT stations to Joo Chiat Road' (Arshad 2006).

The Kampong Glam organisers' plan was a full-scale integrated effort to go beyond the selling of food to break the Ramadan fast in an attempt to relocate the Malay-Muslim's Ramadan attentions, actions and interest to Kampong Glam, which was in the Central Business District, and with it presumably the reinvention or recreation of new memories for the community in a historically Muslim district. Apart from wanting to bring the *Hari Raya* 'mood to the town', it also hoped to draw the 'non-Malay crowd' (Arshad 2006).[22] It was thus created not only for the Malay-Muslim community's interests and for other Singaporeans, but also presumably tourist traffic, though the latter had to be 'attracted' enough to pay a visit. It was this delicate balancing act of attempting to cater to two fairly disparate communities' needs and interests in one particular month that proved to be most challenging.

The crucial economic base of the Malay-Muslim visitors did not find the activities arranged by the Kampong Glam organisers such as the fair and the concert particularly appealing. While the goods at the fair were considered not especially useful for the celebration of *Hari Raya*, there was also disquiet over

---

22 The organisers of the Kampong Glam celebrations included MegaXpress, MediaCorp TV 12, Suria television channel and Majlis Pusat, which is an umbrella body for Malay-Muslim cultural groups. It was also supported by the Singapore Tourism Board.

the mega concert held to welcome the holy month of Ramadan, within hearing distance of the Sultan Mosque and its congregants. The glitzy concert televised live with contemporary dances and popular songs was deemed inappropriate as private conversations remarked on the inappropriateness of the event and the unintended message it might convey. For some, this seems to be an undesirable form of commodifying their faith and culture. While the Malay-Muslim culture prizes warm hospitality and respect for visitors to their home and space, there are cultural sensibilities that must be observed. As noted by Brian J. Shaw: 'Most importantly, the Southeast Asian cities' legacies of ethnic pluralism added authenticity to festivals, food promotions and multicultural shows, many being largely staged events that pandered to a growing tourism market' (Shaw 2006, forthcoming). Pandering might have not be the intention at this event, but it was not difficult to construe the event as attempting to do just that.

Essentially, the Kampong Glam move was an attempt to seize an opportunity, but it was also a miscalculation of what engages the Malay-Muslim community. The community already has strong emotional, historical and spiritual ties with Kampong Glam due to the presence of a major mosque, the Masjid Sultan [Sultan Mosque] and other related Muslim businesses and establishments. This continues in complex and visible ways: for example, the Friday prayers, the shops catering to religious needs, the legendary popular Indian Muslim restaurants and other Malay-Muslim establishments.

However, there are inherent contradictions and difficulties within a rather ambitious Ramadan plan to redirect the larger Malay-Muslim population to an area that is also becoming increasingly alienating to them and in which it is increasingly difficult for them to feel comfortable or affirmed. There are alcohol-selling establishments on the historic Bussorah Street framing the mosque and recent reports of sex workers becoming more visible in the core district of Kampong Glam itself (Mohamed Yusof 2007). As an attempt to redirect economic behaviour and maybe reinvent new traditions and memories, the experiment exacerbated the already troubled reflection on the direction of Kampong Glam and some are definitely not reassured or reaffirmed. Not only has the cultivated glamorous image for the young and trendy and the tourists amplified for some members of the community how globalisation has relentlessly commodified their culture, faith and space with its attending sense of siege, but there is now a need for greater vigilance on the remaking of Geylang Serai's identity.[23]

The English and Malay press gave their unequivocal verdict on the challenge to Geylang Serai: 'It's crowded in Geylang Serai, while Kampong Glam is deserted' (Arshad 2006) and 'Niaga di Geylang Serai lebih rancak dari Kg Glam' (Hamzah

---

23   The establishment of a pub in the former Changi cinema (a building mentioned in the song, 'Di waktu petang di Geylang Serai') in the heart of Geylang Serai did not endear many hearts with fears that it might be the start of attracting 'undesirable' trade and elements into the area.

2006b) ['Business in Geylang Serai is livelier than Kampong Glam'].[24] It was reported that more than 1.7 million visitors came to Geylang Serai with Kampong Glam pulling in only 325,000. However, of these, approximately 30,000 were tourists to Geylang Serai as opposed to the 40,000 to Kampong Glam (Eunos 2007). Suffice it to say, the next Ramadan, in 2007, in Kampong Glam was low-key but still marked by lights, the seasonal food bazaar in Kandahar Street and sales in the shops, but there was no tented bazaar or mega concert. On the other contending space, the celebrations in Geylang Serai have assumed a life of their own and progressively morphed into a bigger, louder and noisier affair in their second year in the temporary location. It is tempting to draw several conclusions from the statistics and couple these with the contentions stated previously, but the fundamental questions in the remaking of space and place for a community are ultimately, for whom are the changes being made, for whose consumption, and at whose expense? (Teo 2003, 547). These are essential questions even for Geylang Serai, whose changes are for redevelopment and not conservation as understood, for example, in the designated conservation areas such as Chinatown and Kampong Glam.

Conclusion: Re-imagining and/or Sustaining Identity?

In the contestation of space, memory and heritage, the dialectic over Geylang Serai can be essentialised to whether it is a space with a community or a community with a space? The National Archives concluded that Geylang Serai is a story of 'Man and his ever changing environment' (National Archives 1986, foreword), but the environment that is sustaining Geylang Serai is a product of state-directed actions, socioeconomic circumstances and power, and, of course, members of the community. In this complex interplay, sustainability will have to incorporate 'factors or components perceived to be the most relevant in a specific context and place that would have the greatest impact on the outcome of sustainability' (Shaw, Wong and Goh 2006, xi). There will be negotiations, compromises and perceived sub-optimal outcomes but the ethical considerations, not always visible or articulated, must be considered. This has made such imperatives crucial especially in considering any minority interests or community with the attending assumptions, perceptions, stereotypes and prejudices. In the essential 'politics of recognition' (Taylor 1994) framework, a 'misrecognition' of interests, aspirations, hopes of the community's identity and sustainability can occur without any apparent sense of awareness or acknowledgement.

---

24   In a 17 October 2006 Radio Singapore International report in Malay, the assessment was that Geylang Serai's primacy 'belum dapat ditandingi' ['has not been matched']. See <http://www.rsi.sg/malay/kpi/view/20061017110124/1/.html> (accessed November 2007).

While Geylang Serai is ostensibly a renewal and upgrading project, an articulated hope has also been the not unwelcomed tourist value it might bring to the area. As hoped by some, the fortunes of the Malay Village could be revived but it would be a bigger economic bonus if the area could become 'good' enough for tourists to come to the off-the-beaten-track Geylang Serai (Anon. 2 Nov. 2006).[25] However, a visiting tourist would come primarily to see not just a Singapore community but an ethnic minority Singapore community 'performing authentically' at being Malay in 'authentic settings'.

While the suggestion might be harmless enough, the concerns for how much the place would have to accommodate or revise its settings, its practices, its environment, its comfort level and at times its economic activities to accommodate the tourist trade might not necessarily stir many hearts. While this might not be the worst case scenario, it must be remembered at whose expense and to what limits this commodification can be encouraged if the decision to pursue the policy is nudged along. A look at glamorous Kampong Glam is a cautionary tale. Can one take heart from the disappointing outcome in Kampong Glam during Ramadam in 2006 that the 'pressure from below', from the people who make Geylang Serai the actual or spiritual home, provides the necessary reality checks to such contentious aspirations? Or from the fact that the *Hari Raya* celebrations and Geylang Serai itself, with its rich history of market traders and residents/visitors, is adequately sustainable with or without the tourists?

Geylang Serai, as with other heartland estates in the country, is where an individual, a family, a community feels comfortable and affirmed. No resident would want to feel that their neighbourhood is a tourist theme park with them as the main attraction: their cultural and social differences not framed as ordinary lives with normal realities and challenges but framed as exotic or 'stylised' (National Archives 1986, foreword) differences with practices that are 'quaint', 'rustic' and 'charming' (whatever that might mean or imagine). While not all tourists or non-Malay Singaporeans might view Geylang Serai or any other ethnic quarter in this romanticised version, a concerted programme to market a heartland neighbourhood as an 'ethnic quarter' will have long-term debilitating social and cultural consequences. Geylang Serai may not be the 'centre of activity' (A. Rahman 2005, 112) for the tourist trade or other non-Malay-Muslim Singapore communities, but it is a thriving 'centre of activity' with engaged involvement by the 'living' and 'moving' people and culture (Mohamed, Ahmad and Ismail 2002, 21): those with valued links to the space, especially the Malay-Muslim community.

---

25   As concluded by Minister Mentor Lee Kuan Yew: 'So we may develop Kampong Glam, the Malay Village in some sense should be transferred and transposed into Kampong Glam. It is out of the way; business is not there because it is not the centre of activity. You can't make a living from Hari Raya as in between the costs are running. So, you build it like a kampong in Kampong Glam – just a small section. I'm sure the tourists will find it attractive. When you rebuild, the large portion of the business comes from tourists.' See A. Rahman 2005, 112.

Ultimately, it is one of the few places where the Malay-Muslim community is not defined by its ethnic difference from the majority ethnic group, the latter usually passing for what is 'normal' or 'mainstream' in Singapore. However, to touristify and commodify the place means to subject the community to definitions of themselves as different and hence not 'normal' in the very area where their numerical and physical visibility seemingly makes their cultural and social practices 'normal'. To re-imagine Geylang Serai as an uninhibited tourist destination is dangerously close to being an extension of the touristification of the 'other' with all its potentially condescending and trivialising fallout.

As it is, Geylang Serai as an 'ethnic quarter' is not simply a site for the intrepid tourists. It is also a site for heritage tours by school parties sometimes in full 'othering' mode or what this author notes as a distressing National Geographic channel approach to 'difference': Singaporeans observing a Singapore ethnic minority as some form of anthropological specimen complete with commentary phrases such as 'these people like ...', 'these people eat ...' and 'these people wear ...'[26] – the reliance on the three 'F's of multiracialism and trivialising the realities of life in a closely managed, highly racialised multiracial society.

Fundamentally, however a space does not have to be classified a heritage site to be a place of heritage significance. As defined by Shaw, 'in the everyday lives of most people, heritage is more about identity and belonging at the local scale, part and parcel of perceived quality of life as familiar landscapes, streetscapes, buildings and activities hold together the fragmented fabric of collective memory (Shaw 2006, forthcoming).

Thus, it is not just a question of visibility or affirmation or that the new market will have 'an architectural style reminiscent of rustic Malay dwellings' (*Kampung Ubi News*, October 2005, 3), but of how one engages with that visibility or supposed affirmation. It is not just about multiracial 'interaction' or 'understanding', no matter how liberally or automated those terms are used in the Singapore context, but ultimately about the meaningful and respectful understanding of differences, taking into account all the subtle or sometimes not so subtle political, economic and numerical dynamics at play. There is a thin line between touristification and affirmation, with questions over who gets to control the representation, the content, the selection and the narratives of the process of transformation. Suffice it to say, Geylang Serai as it is being reconstructed now has displayed the level of recognition that has marked the country's urban renewal and resettlement programme of the past decades.

While one might call it sentimental in wanting to '*balik kampong*' in 2010, it is part of the Malay-Muslim identity that has its roots in Islamic faith, family, culture and home. As noted by the popular traditional Malay folksong, it is about '*pulang bersama-sama*' ['returning together'] to Geylang:

---

26 Anecdotes shared by respondents and interviewees during official duties: workshops, lectures and research on Multicultural issues in Singapore.

Geylang, sipaku Geylang, Geylang, si rama rama.
Pulang, marilah pulang, marilah pulang bersama-sama.
Pulang, marilah pulang, marilah pulang bersama-sama.[27]

References

A. Rahman, Sa'at (ed.) (2005), *The Heart of Geylang Serai* (Singapore: Kampong Ubi Citizens' Consultative Committee).

Ismail, R. (2007), 'Muslims in Singapore as a Case Study for Understanding Inclusion/Exclusion Phenomenon', Paper presented at the Fulbright Symposium 2007: 'Muslim Citizens in the West: Promoting Social Inclusion', Centre for Muslim States and Societies, The University of Western Australia, Perth, Australia, 2 August 2007 (forthcoming).

Mydin, I. (2006) 'The Singapore Malay/Muslim Community: Nucleus of Modernity', in Khoo, Kay Kim, Abdullah, Elinah and Wan, Meng Hao (eds), *Malay Muslims in Singapore: Selected Readings in History 1819–1965* (Malaysia: Pelanduk Publications and RIMA).

National Archives (1986), *Geylang Serai: Down Memory Lane: Kenangan Abadi* (Singapore: Heinemann Asia).

Ooi, G.L. and Shaw, B.J. (2004), *Beyond the Port City: Development and Identity in 21st Century Singapore* (Singapore: Pearson Prentice Hall).

Powell, R. (1997), 'Erasing Memory, Inventing Tradition, Rewriting History: Planning as a Tool of Ideology', in Shaw, B.J. and Jones, R. (eds), *Contested Urban Heritage: Voices From the Periphery* (Aldershot: Ashgate).

Savage, V.R. (1992), 'Landscape Change: From Kampung to Global City', in Gupta, A. and Pitts, J. (eds), *The Singapore Story: Physical Adjustments in a Changing Landscape* (Singapore: Singapore University Press).

Shaw, B.J. (2006), 'Historic Port Cities: Issues of Heritage, Politics and Identity', Paper presented at the SEAGA 2006 Conference, Singapore, November 2006 (forthcoming).

Shaw, B.J. and Ismail, R. (2006a), 'Housing Singapore's Malay Minority', in Goh, Kim-Chuan and Sekson, Yongvanit (eds), *Change and Development in Southeast Asia in an Era of Globalisation* (Singapore: Pearson).

Shaw, B.J., Jones, R. and Ooi, G.L. (1997), 'Urban Heritage, Development and Tourism in Southeast Asian Cities: a Contestation Continuum' in Shaw, B.J. and Jones, R. (eds), *Contested Urban Heritage: Voices from the Periphery* (Aldershot: Ashgate).

Shaw, B.J., Wong, T.C. and Goh, K.-C. (2006), 'Introduction: Urban Sustainability Challenges in an Environment of Integrative Approach' in Wong, T.C., Shaw,

---

27 In a series of rhyming couplets best sung in groups, the song essentially urges returning together … to Geylang Serai.

B.J. and Goh, K.C. (eds), *Challenging Sustainability: Urban Development and Change in Southeast Asia* (Singapore: Marshall Cavendish).

Taylor, C. (1994), 'The Politics of Recognition', in Gutman, Amy (ed.), *Multiculturalism* (Princeton, NJ: Princeton University Press).

Urban Redevelopment Authority (1994), *Geylang Planning Area: Planning Report 1994* (Singapore: Urban Redevelopment Authority).

Journal Articles

Alonso, A.M. (1994), 'The Politics of Space, Time and Substance: State Formation, Nationalism and Ethnicity', *Annual Review of Anthropology* 23, 379–405.

Chang, T.C. (1997), 'Heritage as a Tourism Commodity: Traversing the Tourist-Local Divide', *Singapore Journal of Tropical Geography* 18:1, 46–68.

Chang, T.C. (2005), 'Place, Memory and Identity: Imagining "New Asia"', *Asia Pacific Viewpoint* 46:3, 247–53.

Chih, H.S. (2002), 'The Quest for a Balanced Ethnic Mix: Singapore's Ethnic Quota Policy Examined', *Urban Studies* 39:8, 1347–74.

Chih, H.S. (2003), 'The Politics of Ethnic Integration in Singapore: Malay "Regrouping" as an Ideological Construct', *International Journal of Urban and Regional Research* 27:3, 527–44.

Ho, K.C. (2006), 'Where Do Community Iconic Structures Fit in a Globalizing City', *City* 10:1, 91–100.

Ismail, R. (2006), 'Seasonal Spaces: Ramadan and Bussorah Street – The Spirit of Place', *GeoJournal* 66:3, 243–56.

Kong, L. and Yeoh, B.S.A. (1994), 'Urban Conservation in Singapore: A Survey of State Policies and Popular Attitudes', *Urban Studies* 31:2, 247–65.

Leifer, M. (1967), 'The British Presence and Commonwealth Rivalry in South-East Asia', *Modern Asian Studies* 1:3, 283–96.

Mohamed, B., Ahmad, A.G. and Ismail, I. (2002), 'Heritage Route Along Ethnic Lines: The Case of Penang', *Historic Environment* 16:2, 18–22.

Mutalib, H. (2005), 'Singapore Muslims: The Quest for Identity in a Modern City-State', *Journal of Muslim Minority Affairs* 25:1, 53–72.

Neville, W. (1966), 'Singapore: Ethnic Diversity and Implication', *Annals of the Association of American Geographers* 56:2, 236–53.

Shaw, B.J. and Ismail, R. (2006b), 'Ethnoscapes, Entertainment and 'Eritage in the Global City: Segmented Spaces in Singapore's Joo Chiat Road', *GeoJournal* 66:3, 187–98.

Sim L.L. (1996), 'Urban Conservation Policy and the Preservation of Historical and Cultural Heritage, *Cities* 13:6, 399–409.

Teo, P. (2003), 'The Limits of Imagineering: A Case Study of Penang', *International Journal of Urban and Regional Research* 1:2, 546–53.

Teo, P. and Yeoh, B.S.A. (1997), 'Remaking Heritage for Tourism', *Annals of Tourism Research* 24:1, 192–213.

Yeoh, B.S.A. and Teo, P. (1996), 'From Tiger Balm Gardens to Dragon World: Philanthropy and Profit in the Making of Singapore's First Cultural Theme Park,' *Geografiska Annaler B* 78, 27–42.

Newspaper Articles

Anon. (2003), 'Mixed Response From Stakeholders on Geylang Serai's Redevelopment', *The Straits Times*, 21 February.
Anon. (2003), 'Feedback From Geylang Serai Shopowners to be Given to HDB', *The Straits Times*, 24 February.
Anon. (2006), 'Geylang Serai Market to be Rebuilt', *The Straits Times*, 27 February.
Anon. (2006), 'Anda Diundang Raikan Perasmian Pasar Sementara Geylang Serai', *Berita Harian*, 7 April.
Anon. (2006), 'Semua Dipersilakan Hadir', *Berita Harian*, 8 April.
Anon. (2006) 'HDB Lancer Projek Perumahan "Sri Geylang Serai"', *Berita Harian*, 27 October.
Anon. (2006), 'Citra Melayu Pada Geylang Serai Yang Lebih Indah', *Berita Harian*, 2 November.
Anon. (2006), 'SM Visit Popular Market', *The Straits Times*, 6 November.
Arshad, A. (2006), 'It's Ccrowded at Geylang Serai, While Kampong Glam is Deserted', *The Straits Times*, 21 October.
Eunos, N. (2007), 'Geylang Serai Bermandikan Cahaya', *Berita Harian*, 9 September.
Hambari, H. (2006), 'Joo Chiat Complex Bakal Berwajah Baru', *Berita Harian*, 17 October.
Hamzah, F. (2006a), 'Surau Dekat Pasar Sementara Geylang', *Berita Harian*, 9 April.
Hamzah, F. (2006b), 'Niaga Di Geylang Serai Lebih Rancak Dari Kg Gelam', *Berita Harian*, 23 October.
Hashim, N. (2006), 'Pasar Lambak Hujung Minggu di Kampung Melayu', *Berita Harian*, 12 May.
Hussaini, C.F. (2006a), 'Tindakan Atas Pekerja Haram', *Berita Harian* 5 October.
Hussaini, C.F. (2006b), 'Lebih 1,800 mohon 447 unit', *Berita Harian*, 18 November.
Hussin, H. (2006) 'Pasar Geylang Serai Tinggal Kenangan', *Berita Harian*, 13 April.
Mohamed Y.Z. (2007), 'Streetwalkers Move into Kampong Glam Area, Members of the Community Say … "Too Close for Comfort"', *The New Paper*, 9 November.
Mohd., A. (2006), 'Joo Chiat Complex Sarang Subutex', *Berita Harian*, 17 May.
Mohd., A. (2006), 'CNB Serbu Joo Chiat Complex', *Berita Harian*, 21 May.

Mohd., A. (2006), 'Sri Geylang Serai Dapat Sambutan Hangat', *Berita Harian*, 31 October.

Omar, S. (2006), 'Cebisan Kenangan Di Flat Lama Dibawa Bersama', *Berita Harian*, 20 October.

Salim, S. (2006), 'Faishal Ddisambut Mesra Pegerai Pasar Geylang', *Berita Harian*, 21 April.

Saparin, N.H. (2006), 'Peniaga "Kecohkan" Joo Chiat Complex', *Berita Harian*, 23 May.

Internet-based References

'Census of Population 2000', *Singapore Department of Statistics, Ministry of Trade and Industry, Republic of Singapore*, 2001 <http://www.singstat.gov.sg/sitemap/sitemap.html>, accessed July 2007.

Kwan, T.Y. 'Singapore's Only Halal Wet Mmarket', *Can Book of Records*, 24 September 2007 <http://www.can.com.sg/neocan/en/streetwise/can_book_of_records/gelang_serai_market.html> accessed November 2007.

'New Site Identified for Selective En Bloc Redevelopment Scheme and Hawker Centre Upgrading Programme at Geylang Serai', News Release No: 6/2003, Issued Jointly by HDB and NEA, *National Environment Agency*, 20 February 2003 <http://app.nea.gov.sg/cms/htdocs/article.asp?pid=1948>, accessed October 2006.

'Speech by Mr Mohamad Maidin Packer Mohd., Senior Parliamentary Secretary, Ministry of Home Affairs and Ministry of the Environment, Advisor to Kampong Ubi Grassroots Organisations at the Launch of Exhibition on "Preliminary Design of New Geylang Serai Market" at Kampong Ubi Community Centre on Saturday, 24 April 2004 at 7:30pm', *Ministry of Environment and Water Resources*, 24 March 2004 <http://app.mewr.gov.sg/press.asp?id=CDS1156>, accessed October 2006.

'Suria Suria Raya Karnival/Golden Landmark: Rancangan Ehwal Semasa', *Radio Singapore International* (published online 17 October 2006) <http://www.rsi.sg/malay/kpi/view/20061017110124/1/.html>, accessed October 2007.

'Uniquely Singapore: Geylang Serai', *Singapore Tourist Board* <http://www.visitsingapore.com/publish/stbportal/en/home/what_to_see/ethnic_quarters/geyland_serai.html>, accessed July 2007.

Urban Redevelopment Authority, *Geylang Planning Area: Planning Report 1994* <http://www.ura.gov.sg/dgp_reports/geylang/main.html>, accessed November 2007.

Chapter 3

Paradise Lost? Islands, Global Tourism and Heritage Erasure in Malaysia and Singapore

Ooi Giok Ling and Brian J. Shaw

Introduction

Geopolitics has brought some islands in the Southeast Asian region into prominence in recent times, like the Spratly Archipelago in the eastern central part of the South China Sea (Valencia 2000), variously claimed by China, Vietnam, Taiwan, the Philippines and Malaysia; or Pedra Branca (Savage and Yeoh 2000) also in the South China Sea, where ownership is the source of a dispute between Singapore and Malaysia. Pedra Branca on which the Horsburgh Lighthouse has been built, forms the eastern entrance to the Singapore Straits. One island at least, that of Sebatik, is divided between Malaysia and Indonesia (Teh 2000). While these disputes concern the identities, or at least the national identities of the islands, they are also the basis of territorial claims that could imply control of enormous areas of maritime space and resources. Historically, islands have been in key locations that have made them geopolitically significant among nation-states. This means that islands have long been contested territories. The discussion which follows intends to highlight yet another aspect of islands as sites of intense contestation in today's globalising world. Such contestation concerns the heritage and cultural identity of these islands and their population at a time when the global tourism industry has been increasingly focusing its attention on them as international travel destinations.

Generally, islands have assumed the image of being geographically distant from the mainland and hence, isolated and relatively different in many aspects including most of all, a slower pace of development and change. Not surprisingly, islands have provided getaways for holidaymakers as well as shelter and rest for wayfarers. The development of tourism on islands can be traced back to the Romans who first developed the Isle of Capri as a holiday destination more than two millennia ago (Conlin and Baum 1995). Islands that are scattered throughout the region of Southeast Asia have served historically as homes to seafaring communities and fishermen. More interestingly, these islands have also served as detention centres for political prisoners as well as for refugees like the Vietnamese 'boat people' in the 1970s and for quarantine purposes. During the British colonial period, the island of Sentosa served as a military base. Many other islands have also been used as bases for places of worship with some continuing to be so used to the present day. Inevitably, islands are known to weekend anglers, fishermen and fishing enthusiasts for being the best grounds for different types of fishing, during different seasons of the year.

Islands have long been of interest to researchers, not least in Southeast Asia because of their rich diversity not only in marine life but also terrestrial plant and animal life. The isolation of many of the islands particularly those which are relatively small, means that the nature to be found on them has almost been pristine as yet untouched by the commercial exploitation that some of the larger islands have seen. 'The uniqueness and fragility of an island is influenced by its size, distance from the mainland, geology and manner of evolution. Low-lying islands are especially vulnerable to erosion and drowning from greenhouse-induced sea level rise. Island development poses special challenges as to whether the proposed activities are compatible with the island ecology' (Teh 2000, 7). In other words, the contestation that islands are seeing concerns not only their future in terms of cultural identity and heritage but also long term sustainability that can be threatened by environmental impacts of the development planned on them.

Islands of Malaysia and Singapore

The western coastal areas of Peninsular Malaysia were once considered the most intensively exploited of British colonies in Southeast Asia. Located within these territories, the three Straits Settlements of the port of Malacca (*Melaka*), and inshore islands Penang (*Pulau Pinang*) and Singapore 'were *in* but not *of* the Malay world that surrounded them' (Osborne 2000, 76), being colonial urban bastions characterised by majority Chinese populations. At the same time, islands further offshore such as those in the Langkawi archipelago remained relatively undeveloped during the colonial period of resource extraction, which gathered pace during the latter half of the nineteenth century. Today, both inshore and offshore islands off the coastline of Peninsular Malaysia face intensive exploitation of their natural and cultural heritage in the face of accelerating economic demands resulting from continued population growth, industrial and urban development and the rapidly growing impacts of global tourism.

Using the analogy of Prospero's Isle,[1] Pico Iyer (Iyer 2000, 112) has likened tourism development in Bali to that of a paradise invaded by aliens. In the same way, recent rapid development of the islands of Penang and Langkawi (*Pulau Langkawi*), in terms of theme and industrial parks, hotels and beach resorts, has been disruptive of these islands' rich cultural legacy. In common with tourism in much less developed countries, the responses of local entrepreneurs to the varied demands of visitors to multicultural districts can, and has, led to dissonant

---

1　Prospero's isle in William Shakespeare's play *The Tempest* was the pristine environment on which the succeeding drama was played out.

landscapes of multiple realities and contested meanings (Tunbridge and Ashworth 1996; Aitchison et al. 2002; Shaw et al. 2004). This chapter considers the implications of the (re)construction of such places in social and physical terms, as place consumption and the contestation over cultural identity create the tensions reflected in the place-narrative of Singapore's Sentosa island and the once pristine islands of Langkawi and Penang.

International Tourism and Island Resorts

The tourism sector ranks as the third largest economic sector in Malaysia, behind the revenues generated by the manufacturing sector and the oil palm industries (Malaysian Tourism Promotion Board 1998). In 1995, before the Asian economic crisis, the earnings of the tourism sector totalled some MR10.5 billion (Malaysian *Ringgit*, equivalent to an estimated US$3 billion). Subsequent economic recovery has doubled the size of the economy since 1995, tourist arrivals reached 17.5 million in 2006, and tourism receipts in 2007 have been estimated at MR44.5 billion (Ministry of Tourism 2007). To a large extent, this rapid growth in Malaysian tourism has relied heavily on the country's natural heritage including the many offshore islands along both the western and eastern coastlines of the peninsula, together with protected nature reserves. The most important island resorts include Langkawi, Penang and Pangkor along the west coast; and Tioman, Redang and the Perhentian Islands along the east coast.

The development of island resorts for tourism has generally tended to follow a rather formulaic process, in part dictated by those government agencies charged with the promotion of the tourism sector. For example, within Malaysia tourism development is under the control of the Ministry of Tourism and the Malaysian Tourism Promotion Board (1992). In Singapore responsibility falls upon the Singapore Tourism Board (STB), first established in 1964 as the Singapore Tourism Promotion Board (STPB). Tourism receipts in the city-state were over S$6.2 billion in 2004, and visitor numbers exceeded 9.7 million in 2006 (STB 2007). While national governments in Southeast Asia are seen to be cooperating in the effort to promote tourism there is also intense competition between countries, as evidenced in the regular promotion of 'Visitor Years' years in which overseas promotion is combined with domestic festivals and events. Thus, 'Visit Thailand Year' in 1987 was followed by 'Visit Malaysia Year' 1990, 'Visit Indonesia Year' 1991 and, somewhat inevitably, 'Visit ASEAN Year 1992' (Shaw 2006). Malaysia is currently extending the successful 'Visit Malaysia 2007' invitation through to 31 August 2008 (Ministry of Tourism 2007). Furthermore, competition is not only intense between coastal and inland locations, and among island resorts located within the region, but increasingly occurs between competing island and marine developments spread throughout the world.

Developments on the island resorts have taken the form of large-scale, capital-intensive projects as well as small-scale ventures. Among the former are the international beach resorts and accompanying jetties, developments that have taken precedence over rural interests with the displacement of entire settlements and the collapse of activities like fishing and farming. These latter activities are gradually abandoned because providing transport services for tourism has proven more lucrative (Voon 2000). In addition, the sidestepping of local cultures has been a familiar and iterative outcome of island resort development (Warren 1998). In Penang, local devotees no longer go to some temples that are on the tourist itineraries because the tourists have 'chased away the 'spiritual presence'. Conservation by the state has made the temples 'too colourful, like a zoo for the gods' (Teo 2003, 558). Indeed, locals are discriminated against at beach resorts because of the cordoning off of beaches for the exclusive use of the hotels that have been developed throughout the islands (Teh 2000; Teo 2003).

In particular, the proliferation of golf courses, a phenomenon that increasingly characterises island resorts, has had similar effects throughout the region. Frequently, local cultures have been disregarded, water supplies have been sequestered and unwilling sellers have been pressured in order to establish the hotel-related courses and country clubs patronised by local elites and visiting tourists (Shaw and Shaw 1999). The developers of tourist attractions on the islands have not worked collectively for the benefit of the islands – as deforestation, air and river as well as water pollution and high bacterial content in water at beaches appear among issues flagged by civil society groups (Teo 2003, 559). While the marketing of the beach resort developments promises sun-kissed beaches and golden sunsets, private sector developers and the state development authorities overseeing tourism have generally neglected the pollution along the beaches as well as the soil erosion at sites abutting these.

Sentosa Island, Singapore

Indeed, the global competition for tourists among island resorts has led to the seemingly endless re-making of the image of islands as tourist destinations, as exemplified by Singapore's Sentosa Island. In the last three to four decades, Sentosa Island has gone through many transformations including the changing of its name. Originally known as *Pulau Belakang Mati* or 'island behind the dead', the name of the island was changed to *Sentosa* or 'Isle of Tranquility' in the Malay language, when it was designated for tourism development in 1972 (see Figures 3.1 and 3.2). Its former identity as a British colonial army base with the only preserved coastal fortification in Singapore – Fort Siloso – has been largely overshadowed by the development of infrastructure and themed attractions aimed at the international tourism market. The fort has been incorporated among these themed attractions and over the years it has also been repeatedly renovated as well as refurbished in keeping with the serial process of makeovers whereby the island resort has been continuously reinvented. Coastal reclamation has extended the area of the island and ferry, cable car ride and a variety of transport services have been added to the

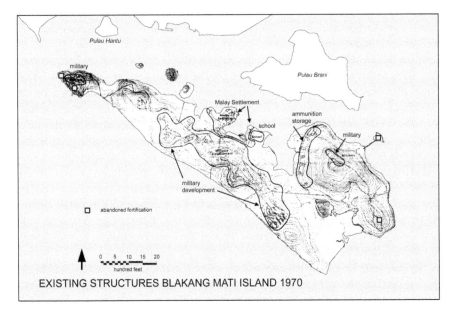

Figure 3.1   Existing structures Blakang Mati Island, 1970

Source: Figure used with permission from *Singapore Architect* with acknowledgement to Mr. Imran Bin Tajudeen.

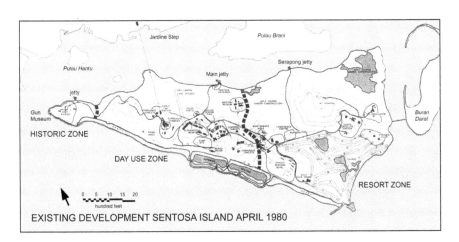

Figure 3.2   Existing development Sentosa Island, April 1980

Source: Figure used with permission from Singapore Architect with acknowledgement to Mr. Imran Bin Tajudeen.

island's infrastructure including a monorail to link the various themed attractions as well as hotel and eating facilities

Sentosa Island started out with the development of a golf course and attractions that were fashioned from existing sites on the island such as Fort Siloso. The initial plan appeared to have been to make Sentosa into an island gateway, a rustic version of what the main island might have been in the early years of economic growth. While tourism development was the goal, the main target would have been day trippers from the Singapore main island – families on outings over the weekends or young party goers looking for places to go to for celebrations and gatherings. Then the recession of the mid-1980s and a fall in tourism arrivals led to the construction of a slew of themed attractions such as *Underwater World*, *Fantasy Island*, *Asian Village* and *Volcanoland*, which appeared during the 1990s. Alongside these were efforts to invent a legend of the origins of the island as well as a variety of themes like the *Ruined City* and the *Lost Civilisation* to add appeal to the attractions being constructed at the time.

In line with Eric Hobsbawm's notion of the 'Invention of Tradition' (1983), the legend created for the island was that of the Kingdom of the Merfolks that has been linked to the STPB's earlier construction of the tourism industry's icon – the Merlion – part lion and part fish (see Figure 3.3). The Merlion was created as a

Figure 3.3   The Merlion at Sentosa

symbol of Singapore that tourists could easily and quickly associate with the city-state, according to the STPB. Part lion because Singapore is the Lion City and part fish in reference to the meeting of land and sea of the island-state, the Merlion is the product of different efforts at creating a tourist symbol for Singapore. Added by the Sentosa Development Corporation in 1993 to create a new storyline for tourists and visitors to the island, the Sentosa Merlion is some 37 metres high or the equivalent of an 11-storey building. Sitting atop a ridge and lit at night it may easily be distinguished by tourists and visitors on the Singapore main island.

On Sentosa Island, 'hardware-driven' developments have been followed up with subsequent demolition as, apart from *Underwater World*, most of the themed attractions experienced falling popularity as similar attractions appeared on the main island itself. In the aftermath of the 1997 Asian financial and then economic crisis, which were associated with a sharp decline in tourism arrivals, the Sentosa Development Corporation proceeded to unveil its 2002–2010 Masterplan aimed at transforming the island resort once again into an upmarket playground for well-heeled tourists, and even residents. In the plan is the development of Sentosa Cove, an exclusive residential area of high-end homes with a 'mega marina' as the lavish centrepiece. A site has also been set aside for the development of an integrated resort complete with a casino and associated shopping, hotel and entertainment facilities. Singapore's former World Trade Centre has been demolished to make way for the development of a new Harbour Front including a mega shopping centre and entertainment facilities. The island resort is now linked to the main island by a light rail facility connecting to Singapore's mass rapid transit system.

For Sentosa Island, tourism development has never considered incorporating aspects of the indigenous seafaring communities of *Orang Laut* who lived in boats in the straits separating Sentosa from the main island. These local residents, the subject of early historical reports compiled in 1819 by the first British Resident William Farquhar, and the significance of the Keppel Harbour Straits, together constitute an important aspect of Singapore's pre-colonial history (Miksic and Low 2004). Neither has there been much attention paid to Sentosa's subsequent role as the island fortress built by the British colonial authorities to protect the harbour and docks that were being developed on the shores of the main island of Singapore. A series of forts and a battery were built on the island to protect what eventually became the second largest dock in the world, the Keppel Harbour trade of the colonial port city of Singapore. This erasure of memory in Singapore has been made possible through the lack of identity in the absence of locally rooted populations, the selective stories of colonial conduct as part of Singapore history, and the efficiency of centralised planning for which Singapore is renowned.

Langkawi Island, Malaysia

In the development of island resorts for global tourism, not only are the local residential population sidelined as well as ultimately marginalised, but virtually much of the history of the islands is generally ignored. The serial re-making of Sentosa Island is not unique in the region and, most certainly, the island resorts of Malaysia have seen a similar process through which tourism promotion authorities have taken these islands in the competition for their share of the international tourism market. The Langkawi archipelago, situated some 30kms off the north western coast of Malaysia but still part of the state of Kedah, has undergone as many, if not more, rounds of image re-making as a tourism destination, as the following discussion will highlight. Observers of the development of coastal resorts in Malaysia such as in Pulau Tioman, point out that 'like other parts of the country, the tourism industry in Tioman Island is largely hardware-driven' (Voon 2000). Emphasis is put on physical development to provide accommodation ranging from the large international resort to village-level chalets, public amenities in transport and communication, water and electricity supplies, recreational space for golfing and swimming, and service-oriented facilities such as restaurants and handicraft shops.

There has similarly, been a general disregard of the history and cultural identity of islands being developed for tourism in Malaysia. Historically, the islands of the Langkawi archipelago, of which there are some 99, have been home to seafarers, pirates and fishermen. However, little of such history or cultural identity has mattered in the development implemented to make the main island of Pulau Langkawi a major tourist destination. Here, development has followed a pattern almost similar to that of Sentosa Island. Initially a golf club was developed with a view to promoting the island as a tourist destination (Teh and Shamsul 2000). This golf course was eventually abandoned because of a lack of funds to maintain the facilities and the lack of interest among tourists as well as the local farming and fishing community. When plans were revived in 1989 for the development of an island resort, the golf course was re-developed together with many themed attractions. Land reclamation was also carried out in order to build an airport (See Figure 3.4).

The earlier series of themed attractions included sites such as the burial grounds of Princess Mahsuri who is the figure at the centre of a legend on which tourist promotional effort has focused. Wrongfully put to death, the princess was believed to have laid a curse on the island of Pulau Langkawi for seven generations. Her story has been woven into a themed attraction that had been highly successful in drawing local tourists to the island resort. Visitors have included Buddhists from neighbouring Thailand who have gone to the burial grounds of the Princess Mahsuri to offer homage for her martyrdom. Historical links between Pulau Langkawi and Thailand appear to have been boosted by such connections and it would seem that the legend is as familiar to Thai visitors as it is to locals. Other themed attractions that were developed in this early phase of tourism development had little to do with the legend and included an underwater world, a handicraft centre, a museum to exhibit the rice-planting culture as well as other local sites related to past conflicts among the people living on the island. The fate of these early themed attractions has mirrored that of similar attractions developed on Sentosa Island and other island resorts. Popularity has declined and many have fallen into disrepair and neglect. This has been followed by a more recent round of construction of themed

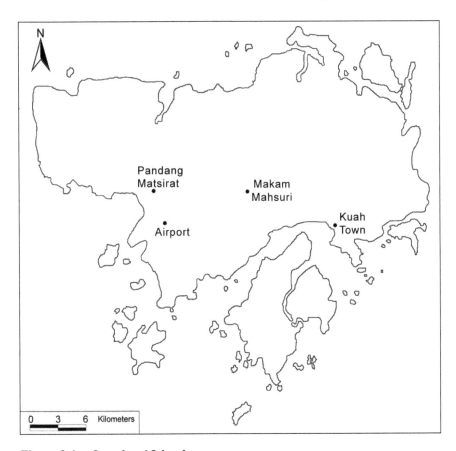

Figure 3.4   Langkawi Island

Source: Figure by author, Ooi Giok Ling with contribution by Chang Chew Hung.

attractions including the development of luxury hotels, which have been allocated coastal sites and beaches in the most attractive parts of the island.

The more recent developments on Pulau Langkawi have included a cable car ride, a Mediterranean-style shopping centre, a tourism complex built around the legendary Princess Mahsuri as well as a new ferry terminal and theme park featuring Malaysian legends in general. Redevelopment of the Mahsuri burial grounds and shrine has involved a private sector firm from the capital city of Kuala Lumpur. In the process, the shrine of Mahsuri has become one of the many attractions offered by the theme park – Kota Mahsuri – that now includes cultural dances, and a small zoo. The complex includes a museum within which the tombstone that had earlier designated the shrine and burial site of the Princess Mahsuri has been consigned. Indeed, the new owners of this themed attraction appear to find the visits by tourists to worship and pay homage to the Princess rather objectionable. This particularly concerns the visits by Buddhist worshippers who often leave incense burning at the site designated as the burial ground. The owners have expressed the wish that the complex become more a tourist destination rather than the shrine to commemorate the martyrdom of the Princess Mahsuri.

The cultural sites on Pulau Langkawi have been re-invented anew to create greater appeal to assist the development of tourism. In the pursuit of their share of the tourism market, the developers of island resorts have largely ignored their local history and cultural legacy. The livelihoods of the local communities like those of Pulau Langkawi have been bypassed and neglected as tour operators strive to enhance the appeal of the island resort to international tourists. While the local populations have had little say in guiding the destiny and imagery of their island homes, they still remain the most consistent and permanent feature of the tourism destination. This situation is somewhat different from that of Sentosa in that a locally rooted population is still largely in existence, but this community has not been co-opted into the development of tourism related imagery, neither has it displayed a sufficient level of resistance to challenge the commercial (re)construction and (re)commodification of place.

Penang Island, Malaysia

Named after the betel-nut palm, the island was occupied by the British in 1786 when Captain Francis Light established a trading station for the East India Company. Although well known to seafarers the island was reported to be almost uninhabited when the British took possession, renaming the settlement George Town and the island Prince of Wales in honour of King George III and his heir. Designed to be a bridgehead for the expansion of British influence in the region, Penang never fulfilled its early promise being overshadowed after 1819 by the rapid development of Singapore. During the colonial period the island assumed regional importance as a source of fine spices and pepper and as an entrepôt following the development of sugar plantations and later tin and rubber on the mainland in Province Wellesley (now Seberang Perai), separated from the island by a short strait, just four kilometres wide at its narrowest (Hoyt 1981). Following independence (1957) and the creation of the Malaysian Federation (1963), Penang lost its free port status which had the effect of further reducing commercial activities within an already stagnating economy manifested by growing unemployment, out migration and a restricted agricultural sector. This situation was countered by the creation of the Penang Development Corporation (PDC), a state authority empowered to manage and develop land and promote industrial development (PDC 1989; Jones and Shaw 1992).

From 1972, the introduction of free trade zones (FTZs), successfully combining foreign capital with cheap local labour, particularly within the electronics/electrical and textile/garment industries, transformed the state's economy. Penang developed a reputation as Malaysia's 'Silicon Valley' and despite an economic downturn in the mid-1980s has successfully made the transition to a manufacturing based economy, the third largest within Malaysian states despite its position as the second smallest state in area, and with a 2006 population of just 1.5 million. At the same time, the PDC was also busy promoting the island as an international tourism destination, marketing its image as 'Pearl of the Orient', a somewhat overused adage within the region and better suited to the island's pristine past. Penang's most attractive beachfront on the northern coastline of Batu Ferringhi was transformed from a collection of small fishing villages, isolated bungalows and 'backpacker' accommodations, into a stretch of international hotels and associated tourism and recreation facilities. This transition was facilitated by the upgrading of the island's airport at Bayan Lepas in the south of the island, as the number of international visitors increased from 39,454 in 1970 to 249,118 in 1981 (Jones and Shaw 1992; Wong 1998).

This juxtaposition of 'Silicon Valley' and 'Pearl of the Orient' largely within the small island area of some 290 square kilometres, brought with it inevitable conflicts, despite an initial spatial separation that placed tourism facilities in the north of the island and industrial development in the south. The gradual degradation of the coastal environment and the need to provide infrastructure for new housing, industry and tourism challenged the pristine littoral setting befitting an 'exotic beach resort' (Jones and Shaw 1992). By the late 1980s the *Penang Guide* admitted that 'the seashore here is pretty but there is almost no surf and the area is of more interest as a place to sit quietly and read or watch the sea than as a place to swim' (Kratoska 1988, 96). In this context, and faced with increasing competition from other developing resort destinations, Penang increasingly marketed itself as a cultural treasure of Malay, Chinese, Indian and other heritages waiting to be discovered through food, festivals and friendly hospitality. Today, this has developed into an industry that complements the agglomeration of Information Technology (IT) in the form of MICE (Meetings, Incentives, Conventions and Exhibitions) events, promoting Penang as a hub of the northern growth triangle (IMT-GT) and with an increasing emphasis on domestic tourism which has some different and distinctive demands when compared to the international market (Teo 2003).

We might infer from this brief sketch that 'Paradise Lost' is a condition that impacted early upon Penang as the demands of global tourism imposed themselves particularly severely upon local communities and the environment. However, the situation here is somewhat different from that of both Sentosa, which lacked a locally rooted population, and Langkawi where the community was largely sidelined within the re-imaging process. Penang's much larger population and distinctive Chinese heritage as one of the three original Straits Settlements, combined with a vibrant NGO sector and various community conservation groups, have challenged the opaque recesses of corporate and government decision-making. In the mid-1980s the locally based Consumer's Association of Penang raised a number of early concerns regarding the social and environmental impacts of tourism in Malaysia in a landmark publication entitled *See the Third World while it lasts*

(Hong 1985). The *Pulau Pinang* magazine has raised heritage awareness both within and outside of the state and the publication of *Penang Hill: The Need to Save our Natural Heritage* was a major contribution by the 'Friends of Penang Hill' (1991). More recently the 'Save Ourselves' tenants' self-help group has been formed to protect the rights of tenants affected by the repeal of the state's Rent Control Act (Teo 2003).

Perhaps the most successful community action was taken by the 'Friends of Penang Hill' in the early 1990s, in response to a detailed development plan for the historic site, which is a series of connecting peaks rising to 830 metres located in the centre of the island and a repository of natural and cultural heritage. The area's significance as a refuge for many rare species of flora and fauna, together with its importance as the oldest colonial hill station developed in Peninsular Malaysia, were threatened by a proposal to redevelop the Hill with two large hotels, a condominium, a futuristic 'Summit Acropolis', a Tiger Hill Adventure Park and other entertainment and leisure facilities. Cable cars and an eventual road would open up access to the summit (Friends of Penang Hill 1991). Such themed attractions and the invention of legends, in a similar manner to that which has been inflicted upon both Sentosa and Langkawi, would have totally destroyed the ambiance of Penang's unique asset. Leading the successful fight against this proposal initiated by the state government the 'Friends' operated as a network of public interest groups and concerned individuals that put forward an alternative proposal and organised a public petition with 30,000 signatories (Jones and Shaw 1992).

Despite this example of a successful resistance movement, the struggle continues. While the traditional features of Penang's northern shores may be irretrievably in thrall to the demands of tourism and recreation; and while Penang Hill has, for the present, escaped the worst manifestations of proposed 'Disneyfication'; we should consider the fate of George Town's built heritage. Notwithstanding the recent (2001) joint nomination of the historic centres of Melaka and Penang for World Heritage Status (UNESCO 2007) the future of George Town remains uncertain as ramifications unfold from the repeal of rent controls and the demands of developers gather pace. George Town is now somewhat precariously placed upon the continuum of urban heritage conservation, situated somewhere between the 'decayed' stage lacking the investment needed for conservation and restoration, and the 'pastiche' stage where heritage has been (re)created not necessarily in the context of historical accuracy (Shaw et al. 1997). Peggy Teo's article, to which frequent reference has been made in this chapter, is entitled 'The Limits of Imagineering'. The questions remain, where are these limits and what will we finally have left?

Conclusion

Island resort developments like those of Sentosa Island, Pulau Langkawi and Pulau Pinang, represent the logic of the touristic marketplace which requires recurrent innovation and re-creation where the 'longevity' and 'cultural' layers of a landscape are constantly being fragmented and reworked in tandem with market forces (Zukin 1991, 27). The serial re-making of themed attractions in these island resorts lack a coherence as far as the cultural identity and history of the islands are concerned since the process is dictated largely by the needs of the global tourism marketplace. These themed attractions feature among others, the ubiquitous underwater world and cable car rides. Where the tour guides have incorporated local cultural sites, these have either been invented by tour promoters or else re-cast and re-packaged by the developers for mass consumption.

In disregarding the history and cultural identity of the islands that are being developed for tourism, the logic of the touristic marketplace contradicts the effort as well as the ceaseless process of projecting the attractions of the islands as tourist destinations. The global logic of the tourism market has generally marginalised the local identities of the islands that are being developed. Development of the varying rounds of themed attractions has been in abstraction from the cultural context against which the identity of an island resort can be understood. Yet clearly this need to understand the meanings of the island's heritage – cultural as well as natural – has not been evident in the headlong rush by tourism authorities to develop island resorts for global tourism. The consequence has therefore been the homogenisation of the developments being pitched at the global tourism market and the clumsy patching over of a cultural heritage that belongs to the islands and their people.

References

Aitchison, C., Macleod, N. and Shaw, S. (2002), *Leisure and Tourism Landscapes: Social and Cultural Geographies* (London: Routledge).

Baudrillard, J. (1983), *Simulations* (New York: Semiotext(e)).

Bauman, Z. (1996), *Intimations of Postmodernity* (London: Routledge).

Boyer. M.C. (1992), 'Cities for sale: merchandising history at South Street Seaport,' in Sorkin (ed.).

Conlin, M. and Baum, T. (eds) (1995), *Island Tourism: An Introduction in Island Tourism: Management Principles and Practice* (Chichester: John Wiley and Sons).

Crook, S., Pakulski, J. and Waters, M. (1992), *Postmodernisation: Changes in Advanced Society* (London: Sage).

Friends of Penang Hill (1991), *Penang Hill: The Need to Save our National Heritage* (Penang: Friends of Penang Hill).

Government of Malaysia (1991), *The Second Outline Perspective Plan 1991–2000* (Kuala Lumpur: National Printing Department).

Greider, W. (1997), *One World Ready or Not* (New York: Simon and Schuster).

Gruffudd, P. (1994), 'Back to the Land: Historiography, Rurality and the Nation in Interwar Wales', *Transactions, Institute of British Geographers* 19, 61–77.

Haralambos, M. and Holborn, M. (2004), *Sociology – Themes and Perspectives* (London: Collins).

Harvey, D. (1989), *The Condition of Postmodernity – An Inquiry into the Origins of Cultural Change* (Oxford: Blackwell).

Harvey, D. (2000), `Time-Space Compression and the Rise of Modernism as a Cultural Force', in Lechner and Boli (eds).

Hassan, S. (2000), `Determinants of Market Competitiveness in an Environmentally Sustainable Tourism Industry', *Journal of Travel Research* 38, 239–45.

Hobsbawm, E. and Ranger, T. (eds) (1983), *The Invention of Tradition* (Cambridge and New York: Cambridge University Press).

Hong, E. (1985), *See the Third World While it Lasts* (Penang: Consumer's Association of Penang).

Hoyt, S.H. (1991), *Old Penang* (Oxford: Oxford University Press).

Huybers, T. and Bennett, J. (2003), `Environmental Management and the Competitiveness of Nature-Based Tourism Destinations', *Environmental and Resource Economics* 24, 213–33.

Hall, C.M. and Page, S. (eds) (2000), *Tourism in South and Southeast Asia: Issues and Cases* (Oxford: Butterworth, Heinemann).

Hirsch, P. and Warren, C. (eds) (1998), *The Politics of Environment in Southeast Asia* (London and New York: Routledge).

Iyer, P. (2000), 'Bali: On Prospero's Isle/The Philippines: Born in the USA' in Lechner and Boli (eds).

Johnson, N. (1995), `Cast in Stone: Monuments, Geography and Nationalism', *Environment and Planning D: Society and Space* 31, 51–65.

Jones, R. and Shaw, B.J. (1992), 'Historic Port Cities of the Indian Ocean Littoral: the Resolution of Planning Conflicts and the Development of a Tourism Resource Potential', Occasional Paper No. 22 (The University of Western Australia: Indian Ocean Centre for Peace Studies).

Kratoska, P.H. (1988), *The Penang Guide* (Singapore: G. Brash).

Langkawi Development Authority (1990), *Langkawi Structure Plan 1985–2005*.

Lechner, F.J. and Boli, J. (eds) (2000), *The Globalization Reader* (Oxford: Blackwell Publishing).

Lee, B.T. (1996), `Emerging Urban Trends and the Globalizing Economy in Malaysia,' in Lo and Yeung (eds).

Lo, F.-C. and Yeung, Y.M. (eds), *Emerging World Cities in Pacific Asia* (Tokyo: United Nations University Press).

Malaysian Tourism Promotion Board (1998), *Malaysian Tourism*, July–August.

Marcuse, P. and van Kempen, R. (eds) (2000), *Globalizing Cities: A New Spatial Order?* (Malden, Mass.: Blackwell).

Mathieson, A. and Wall, G. (1982), *Tourism: Economic, Physical and Social Impacts* (London: Longman).

Miksic, J.N. and Low, C-A. (eds) (2004), *Early Singapore 1300s–1819* (Singapore: Singapore History Museum).

Mihalic, T. (1999), 'Environmental Management of a Tourist Destination: A factor of Tourism competitiveness', *Tourism Management* 21, 65–78.

Milne, S. (1990), 'The Economic Impact of Tourism in Tonga', *Pacific Viewpoint* 31(1), 24–43.

Orbasli, A. (2000), *Tourists in Historic Towns: Urban Conservation and Heritage Management* (London: Spon).

Osborne, M. (2000), *Southeast Asia: An Introductory History* (St Leonards: Allen and Unwin).

Peck, J. and Tickell, A. (1994), 'Jungle Law Breaks Out: Neoliberalism and Global Local Disorder', *Area* 25, 317–26.

Penang Development Corporation (PDC) Report (1989).

Ritzer, G. (1996), *The McDonaldization of Society* (London: Pine Forge Press).

Savage, V., Kong, L. and Neville, W. (eds) (1998), *The Naga Awakens: Growth and Change in Southeast Asia* (Singapore: Times Academic Press).

Savage, V.R. and Yeoh, B.S.A. (2003), *Toponymics – A Study of Singapore Street Names* (Singapore: Eastern Universities Press).

Shamsul B. and Teh, T.S. (2000), 'Island Golf Course Development in Malaysia: An Environmental Appraisal' in Teh (ed.).

Shaw, B.J. (2006), 'Urban Heritage Conservation and Tourism Development in Southeast Asia' in Wong, Shaw and Goh (eds).

Shaw, B.J., Jones, R. and Ooi, G.-L. (1997), 'Urban Heritage, Development and Tourism in Southeast Asian Cities a Contestation Continuum', in Shaw and Jones (eds).

Shaw, B.J. and Shaw, G. (1999), '"Sun, Sand and Sales" Enclave Tourism and Local Entrepreneurship in Indonesia', *Current Issues in Tourism* 2, 68–81.

Shaw, B.J. and R. Jones (eds) (1997), *Contested Urban Heritage Voices from the Periphery* (Aldershot: Ashgate).

Shaw, S., Bagwell, S. and Karmowska, J. (2004), 'Ethnoscapes as Spectacle: Reimaging Multicultural Districts as New Destinations for Leisure and Tourism Consumption', *Urban Studies* 41:10, 1983–2000.

Sorkin, M. (ed.) (1992), *Variations on a Theme Park: The New American City and the End of Public Space* (New York: Hill and Wang).

Swarbrooke, J. (2002), *The Development and Management of Visitor Attractions* (London: Butterworth-Heinemann).

Tan, P.K. (1992), 'Tourism in Penang: Its Impacts and Implications', in Voon and Tunku Shamsul Bahrin (eds).

Tan, A.T.H. and Boutin, J.D.K. (2000), *Non-Traditional Security Issues in Southeast Asia* (Singapore: Institute of Defence and Strategic Studies and Select Publishing).

Tan, W.H. (2000), 'Island Tourism Development: A Case Study of the Perhentian Islands, Terengganu,' in Teh (ed.).

Teh, T.S. (ed.) (2000), *Islands of Malaysia – Issues and Challenges* (IRPAR and D Project, University of Malaya, Kuala Lumpur).

Teh, T.S. and Ong, A. (2000), 'An Analysis of Construction Setback Lines of Beach Resorts: With Special Reference to Langkawi Island', in Teh (ed.).

Teh, T.S., Tengku Shamsul Bahrin and Ooi, C.H. (2000), 'Future of Pulau Layang-Layang: Man-Made Paradise or Ecological Disaster?' in Teh (ed.).

Teo, P. (2003), 'The Limits of Imagineering: a Case Study of Penang', *International Journal of Urban and Regional Research* 27:3, 545–63.

Tunbridge, J. and Ashworth, G. (1996), D*issonant Heritage: The Management of the Past as a Resource in Conflict* (Chichester: John Wiley and Sons).

Valencia, M.J. (2000), 'Building Confidence and Security in the South China Sea: The Way Forward' in Tan and Boutin (eds).

Voon, P.K. (2000), 'Tourism and the Environment: The Case of Tioman Island, Malaysia' in Teh (ed.).

Voon, P.K. and Tunku Shamsul Bahrin (eds) (1992), *The View From Within – Geographical Essays on Malaysia and Southeast Asia, Kuala Lumpur* (Malaysia: The Malayan Journal of Tropical Geography).

Warren, C. (1998), 'Tanah Lot: the Cultural and Environmental Politics of Resort Development in Bali' in Hirsch and Warren (eds).

Williams, G. (2003), *The Enterprising City Centre: Manchester's Development Challenge* (London: Spon).

Wong, P.P. (1998), 'Beach Resorts: the Southeast Asian Experience' in Savage, Kong. and Neville (eds).

Wong, T.C., Shaw, B.J. and Goh, K.C. (eds) (2006), *Challenging Sustainability: Urban Development and Change in Southeast Asia* (Singapore: Marshall Cavendish).

Zukin, S. (1991), *Landscape of Power: From Detroit to Disney World* (Berkeley and Los Angeles: University of California Press).

Internet-based References

Ministry of Tourism, Malaysia (*Kementerian Pelancongan Malaysia*) (2007), [website] <http://www.motour.gov.my/news/index.php?/archives/501-Malaysias-Tourism-Revenue-Expected-To-Hit-RM44.5-Billion-This-Year.html> accessed December 2007.

Singapore Tourism Board (STB) (2007), *Key Statistical Information* [website] <http://app.stb.gov.sg/asp/tou/tou02.asp#VS> accessed December 2007.

UNESCO (2007) *World Heritage: Historic Centres of Melaka and Penang* <http://whc.unesco.org/en/tentativelists/1519/> accessed on December 2007.

Chapter 4

'Being Rooted and Living Globally': Singapore's Educational Reform as Post-developmental Governance

Mark Baildon

Managing Globalisation

Globalisation has necessitated new arrangements between the state, markets and populations. Fast capitalism (Agger 1989, 2004; Holmes 2000), transnational flows of people, ideas, goods, media and technologies, and the shifting landscape of the knowledge economy require responsive populations and policies that can flexibly and quickly adapt to ever changing circumstances. In particular, processes of globalisation have 'create[d] a need for modern states to alter their governance modes and to institute reforms in the public sector to better manage uncertainty and risk' (Gopinathan 2007, 54). Educational reform and policy play a central role in mediating and managing these shifting relationships between state and society and are one way governments try to address anxieties accompanying fundamental changes due to globalisation (Koh 2004).

Since new social, organisational and educational strategies are necessary to accommodate and manage the myriad changes accompanying globalisation, 'the reform rhetoric is remarkably similar across very different education jurisdictions; all reform proposals stress the need for greater attention to processes, higher order thinking skills, better utilisation of technology in education, changes to assessment, greater devolution of power to principals, etc.' (Gopinathan 2007, 56). Such reforms can be seen as efforts by states to produce and manage educational systems and populations that are attractive to the shifting requisites of global capital (Ong 1999). As a result, educational reform becomes ever more critical in the never-ending drive for competitive advantage in the global innovation economy.

Singapore, in particular, presents an interesting case of how education is used to promote and sustain economic development in a relentlessly changing and increasingly competitive global economy. Since its independence in 1965, Singapore has implemented educational policy and reform for nation-building purposes, to engineer consensus among diverse ethnic groups, and to promote economic growth and development. In this chapter, however, I argue that the recent suite of educational reforms initiated by Singapore in 1997 under the umbrella term 'Thinking Schools, Learning Nation' represents an attempt by Singapore as a post-developmental state (Ong 1999) to re-calibrate schooling in order to accommodate and manage the new imperatives and potential consequences of globalisation. Instead of seeing recent educational reform in Singapore as characteristic of the developmental or strong state as it has been viewed by several commentators (Gopinathan 1994, 2007; Green 1997; Koh 2002; Sim and Print 2005), it is more useful to see Singapore as a post-developmental state formulating new, more flexible responses to the trajectories of newly perceived global imperatives and local 'sociopolitical and cultural-ideological needs' (Koh 2004).

In particular, the shift from a developmental to a post-developmental state is necessitated by shifts from a manufacturing economy emphasising the standardisation of goods and services and centralised state economic management to a knowledge-based innovation economy that emphasises greater decentralisation, diversification, and local autonomy and initiative. It also entails a shift from strong state efforts to socialise and educate obedient populations necessary for state building toward greater use of cultural and symbolic forms of control that place greater emphasis on self-regulation aligned with a broader portfolio of state interests. In response, state policies, rather than mandating prescribed sets of rules and regulations, focus more on the habitus (Bourdieu 1977) of populations by addressing education and training as well as a variety of cultural and institutional conditions, such as health, communication, leisure and recreation, and sociability (Castells 1999). The aims of the post-developmental state are to produce flexible, responsive populations able to constantly redefine themselves in accordance with continual innovation and change in all spheres of life and the ever-shifting needs of the state and economy.

In this chapter I describe how recent Singaporean educational reform has addressed the challenges of globalisation and governance in this new globalised political economy. I begin by describing 'Thinking Schools, Learning Nation' (TSLN) reforms and then demonstrate how these are efforts by the post-developmental state to manage several key tensions, paradoxes, and contradictions accompanying globalisation and governance in 'new times.' I conclude by identifying several lingering challenges for Singapore's post-developmental education system.

Singapore's Curriculum Imagination

The implementation of several educational reforms initiated in 1997 under the theme of 'Thinking Schools, Learning Nation' (TSLN) had five main thrusts: national citizenship education, greater autonomy and authority for local school leaders, increased use of information technology, emphasis on creative and critical thinking skills, and greater options and choice for stakeholders. TSLN officially consisted of several initiatives from 1997–2004 that included an emphasis on 'Ability-Driven Education' (ADE), an 'Information Technology Masterplan', 'Innovation and Enterprise' initiatives, marketisation and diversification of the upper secondary landscape, and 'National Education' (NE). TSLN's overall goals were to prepare students so they would be 'continually prepared for the future' (Ng 2005, 1).

In 1997, the then Prime Minister Goh Chok Tong launched TSLN by calling for an educational system that would 'better develop creative thinking skills and learning skills required for the future ... bring about a spirit of innovation, of learning by doing, of everyone, each at his own level, all the time asking how he can do his job better.' With an eye toward the future, TSLN set forth a plan for lifelong learning. These national learning goals were prefaced by Goh declaring that a nation's wealth would increasingly depend on the capacity of its people to learn, change, and innovate (Goh 1997). The TSLN plan called for a fundamental review of curriculum and the assessment system. Principals were encouraged to 'think of themselves as CEO's of their schools, and to manage their schools like companies – by leading people, producing results and answering to "shareholders" and "customers", and talking about service, marketing, getting results, bottom-line and vision statements' (Ng 2005, 4).

ADE brought about a third phase in Singapore's educational development. As S. Gopinathan pointed out, 'Education reform in Singapore is primarily a way of retooling the productive capacity of the system' (Gopinathan 2007, 59) and the third phase marked an attempt to retool by way of creating new sets of skills and capacities for globalisation and the information age, commitments to lifelong learning, and greater flexibility and adaptability. The first phase of Singapore's educational system from 1959 to 1978 was a survival-driven education that emphasised technical skills necessary for manufacturing and social cohesion. The second phase, the New Education Policy (NEP), from 1979–1996 was efficiency-driven and based on a model of economic rationality and standardisation to develop human capital for economic productivity. The new ability-driven education instituted by TSLN in 1997 was process-oriented and focused on a reduction of subject content, revision of assessment modes, holistic education, character building and the development of a broader spectrum of talents and abilities. No ability was to be left behind; all talents were to be marshaled for full use by the nation.

In 1997, the 'Information Technology Masterplan' was launched by Rear Admiral Teo Chee Hean, then Minister for Education in a speech that outlined four main goals to: enhance linkages between schools and the world, encourage creative thinking and lifelong learning, stimulate innovating processes, and promote administrative and managerial excellence. Teo noted that Singaporeans needed to 'learn to think beyond the obvious, to think creatively, to search for new knowledge, to come up with new ideas. They must be comfortable with new technologies and be able to exploit these new technologies to venture beyond their current boundaries and open up new frontiers of knowledge.' The main goal of the initiative was to develop competencies required by the information age – the ability to access, analyse, and apply information, learn independently, and use IT in effective and innovative ways (Teo 1997).

By 2004, the succeeding Education Minister Tharman Shanmugaratnam described 'Innovation and Enterprise' as a journey that aimed to develop strength of character, intellectual curiosity, the courage to live with ambiguity, a sense of teamwork and a spirit of inquiry. He referred to these as 'intangible factors' making up a mindset that would 'allow Singapore to stay relevant' (Tharman 2004). Schools were encouraged to innovate and diversify by customising programs that were more holistic and met the talents, abilities and needs of their students. Awards were given to schools based on broad criteria, such as 'value-addedness, best organisational practices, and achievements in sports and the arts' (Ng 2005, 47).

Increasing marketisation and diversification of the upper secondary landscape also accompanied TSLN to provide greater flexibility and choice in education. More flexible admission systems to secondary schools, Junior Colleges, and universities were devised, the school ranking system was revised, and Institutes of Technical Education provided more opportunities for the technical stream of secondary education. Integrated Programmes were introduced so that top students would not be required to take GCE O-Level exams and could move through to GCE A-Levels. Specialised independent schools focusing on sports, maths and sciences, and the arts also opened and privately funded schools were allowed to open to provide further choice to students and parents.

NE arose in response to concerns by the founding generation that young Singaporeans lacked basic understanding about Singapore's history, its constraints and vulnerabilities, and the values and strategies that contributed to nation building. According to Prime Minister Goh Chok Tong, Singapore could not 'afford to have a new generation grow up ignorant of the basic facts of how we became a nation, and the principles of meritocracy and multi-racialism which underpin our entire society and political culture.' He called for the revision of Social Studies, Civics and Moral Education, and History 'to emphasise nation-building [and] engender a shared sense of nationhood.' He stressed that NE must appeal to both heart and mind 'through the discipline and rituals of school life' and cited the examples of Sri Lanka and Northern Ireland as examples of what can happen to Singapore if it did not strive for harmony and prosperity and develop a set of shared values (Goh 1996).

At the launch of NE the then Deputy Minister Lee Hsien Loong further argued that ignorance of the recent past would hinder efforts to develop a shared sense of nationhood and 'maintain the will to survive and prosper in an uncertain world.' He called for systematically transmitting national instincts and attitudes as part of the 'cultural DNA' of Singaporean identity. The objectives of NE were to develop national cohesion, foster a sense of national pride, learn 'the Singapore story' (mainly the hardships and sacrifices of the founding generation and ruling People's Action Party (PAP)), understand Singapore's unique challenges, constraints, and vulnerabilities, and instil the core values of meritocracy, harmony, and good governance. Nothing less than national survival was at stake as Lee concluded his speech by stressing to teachers that …

... moulding the next generation is in your hands. You must imbue them with a strong sense of national identity and social responsibility. If we fail, all that we have painstakingly built up over decades can unravel and fall apart within a few years (1997).

As part of the National Education effort, Social Studies was introduced as a compulsory and examinable secondary subject in 2001 to promote the idea of 'being rooted and living globally.' The examples of Sri Lanka and Northern Ireland were included in the secondary Social Studies syllabus to emphasise the need for multiracial harmony, the importance of a strong national identity, and the development of skills necessary for living in a diverse global society. The future-oriented aims of the secondary Social Studies syllabus were to enable students to:

1. understand the issues that affect the socio-economic development, the governance and the future of Singapore;
2. learn from experiences of other countries to build and sustain a politically viable, socially cohesive and economically vibrant Singapore;
3. develop citizens who have empathy towards others and who will participate responsibly and sensibly in a multi-ethnic, multi-cultural and multi-religious society;
4. have a deep sense of shared destiny and national identity (Ministry of Education 2001).

The restructuring of Singapore's educational system enacted by TSLN reforms was aimed at reducing curriculum by 30 per cent across all subjects so schools and teachers could focus on critical thinking, independent learning, and innovative pedagogy. It sought to provide a more holistic, flexible, and diverse educational landscape so that all talents could be nurtured and eventually utilised in accordance with the needs of the nation. TSLN sought to create a multi-faceted 'total learning environment' so that the entire national agenda was to be aligned with these new learning objectives and provide fundamental support for ongoing learning (Khong 2004).

Global and Local Imperatives

Several speeches by government officials to support NE and the broader TSLN framework consistently placed Singapore's educational reform in the contexts of global, regional, and local imperatives. For example, Deputy Prime Minister Lee Hsien Loong in 1997 noted the dangers of Chinese chauvinism and racial politics in Singapore's elections, Indonesian haze and its effect on Singaporean air quality, and financial instability in Thailand, which had both regional and global repercussions. He stressed that the next century would bring a whole new set of problems and circumstances and that TSLN needed to prepare citizens to be ever-ready to adapt to new realities and problems (Lee 1997). At his 1997 National Day Rally speech, Prime Minister Goh cited the increasingly competitive global environment and how globalisation and technology were transforming the nature of work and communities (Goh 1997). Dr. Aline Wong, Senior Minister of State for Education made a case for Social Studies as a means to develop a deep sense of belonging to community and nation in the face of rapid globalisation (Wong 2000). As A. Koh argues, the re-alignment of the educational landscape can be seen as 'a response to the trajectories of (global) economic conditions, concomitantly framed by (local) sociopolitical and cultural-ideological needs' (Koh 2004, 335).

The perceived trajectories of global economic conditions included rapid technological innovation, increasing competitiveness, a knowledge-based economy requiring greater innovation, creativity, and collaboration along with the ability to work critically with various types of information, and the likelihood of transnational problems that could affect Singapore in unpredictable ways. Rapid change and global flows of goods, ideas, technologies and people would undoubtedly continue to bring tremendous benefits to Singapore's economy but the interdependent nature of globalisation would also bring unprecedented levels of vulnerability that could be difficult to manage.

The asymmetrical, unpredictable, and fragmentary consequences of global capitalism (Brown and Lauder 2001; Stiglitz 2006; Tan and Gopinathan 2000) require greater diversification and fluidity and the capacity for flexible response, continual adaptation and differentiation. In response, recent educational reform includes greater diversification of the educational landscape (more choices available to stakeholders), an increased emphasis on continual learning and adaptation, and greater importance assigned to the ability of the system and individuals to innovate and change.

Recent reform in Singapore adopts a multi-faceted, flexible, almost all-encompassing approach that draws on a diverse range of strategies. National survival itself now rests on continual, non-stop education, as Hawazi Daipi, Senior Parliamentary Secretary in the ministries of Education and Manpower noted in a forum on 'Achieving a Social and Moral Balance in Globalization' in 2005. Daipi argued that Singapore could not survive and prosper if Singaporeans didn't 'make the effort to re-look, re-invent and re-vitalise' themselves. He called on all Singaporeans to ...

> continue to upgrade [their] skills and capabilities, and keep an open mind. Indeed, those who are able to survive and thrive in this new borderless, global environment are those who can respond quickly to take advantage of the new opportunities and meet the challenges presented by globalization (2005).

However, the greatest imperatives driving TSLN reforms were related to identity formation. Schools in Singapore have traditionally played a twofold role in the developmental state: 'to provide students with the skills required in an industrialising and modern Singapore; and to inculcate in them values that will ensure their loyalty and commitment to the nation' (Quah 2000, 78). However, what marks new educational reform is the explicit recognition that globalisation and the changing economy 'will strain the loyalties and attachments of young Singaporeans' (Gopinathan 2007, 61). Due to multiple global flows and new 'mediascapes', 'technoscapes', and 'ideoscapes' (Appadurai 1996), young people are increasingly pulled into multiple allegiances that challenge the hold of the nation state. Identities are less stable. The imagined worlds of 'the official mind and of the entrepreneurial mentality' (Appadurai 1996, 33) are contested and subverted by the range of options available to young people. Identities are increasingly up for grabs in 'new times' (Koh 2004).

While fostering a sense of a national identity with shared values and commitments obviously still remains important to Singapore's government and is reflected in NE policy, there is a concomitant emphasis on cultivating the dispositions of creativity, innovativeness and adaptability. The educated person in 'new times' is the 'portfolio person' (Gee 2000), the person who has a range of abilities and skills, which can be rearranged to meet the demands of any project. They are able to think critically and creatively, work collaboratively, and carry out their responsibilities with minimal supervision. They respond favorably to pastoral inducements to continually upgrade their skills and knowledge and further enhance their portfolio, which is remarkably aligned with the goals, values, and commitments of the state and economy. In the new learning environment, learning is not judged so much by an increase in knowledge, the traditional school measure, but by 'changing participation in changing practices' (Lave 1996, 161). Since practices are continually changing, economic survival for the nation and the individual now requires continual learning, anticipation of the future, and the ability to flexibly adapt to constant change and monitor oneself in new environments.

Post-developmental Governmentality

More specifically, we can see this re-alignment of identity and the educational system as a shift toward post-developmental governance in which schooling becomes a means for managing national populations in response to global and local imperatives. As A. Luke argues, 'all curriculum narrates, projects, "trajects" imagined human subjects into future pathways' (2002, 3) and in the era of intensified global flows these future pathways are aligned with the needs of chaotic global capital, always shape-shifting, following the flows of profit, quick to adjust to new global conditions and imperatives. To align subjectivities with the needs of the post-developmental state, national educational systems represent two basic functions: the attraction function to create human capital that attracts flows of capital investment, businesses, and economic development; and an amelioration function which mops up and manages the unequal distribution of capital and other problems, such as environmental problems and social issues that accompany rapid development and ever changing local landscapes (Luke 2002). This requires a flexible, adaptive, and responsive educational system, just like it requires subjectivities that are flexible, adaptive and responsive, prepared to quickly adjust and learn what is required. In this sense, there is a merger of the interests, values and commitments of the individual and the state. More importantly, individuals must learn to monitor and assess themselves to ensure ongoing productivity, self-discipline and re-calibration, if necessary.

According to A. Ong, this is a hallmark of post-developmental governance. Post-developmental states enact a range of strategies especially designed to produce and manage middle classes and form links with global capital. They emphasise pastoral care to nurture citizens attractive to capital: 'Post-developmental strategies emphasise the caring aspects of state power; those aspects are directed toward the middle class because it is the middle-class citizenry whose credentials, skills, and overall well-being have been so critical to attracting foreign capital' (Ong 1999, 201). Typically, pastoral inducements are communicated through cultural-symbolic discourses rather than more authoritarian forms of social engineering. For example, Singapore utilises the neo-liberal logic of markets and competition and Asian values, such as the ideals of meritocracy and Confucian values of hard work and harmony, to discipline individuals and manage populations. Greater market choice for stakeholders and the encouragement of school administrators to consider themselves as CEO's combined with the rhetoric of meritocracy and ability-driven education are key strategies in TSLN reforms.

The norms, values, and practices suggested by neo-liberalism and Asian values enable individuals to monitor and audit themselves so they 'can optimise choices, efficiency, and competitiveness in turbulent market conditions' (Ong 2006, 6). The personal attributes and dispositions that make up subjectivities in 'new times' can now be monitored and assessed, not only by others but by each individual. It turns the locus of discipline from the government onto individuals themselves so they discipline themselves in ways consistent with the needs of society. Individuals are given greater freedom to create, innovate and think critically but are expected to demonstrate greater self-discipline and self-motivation in ways that are aligned with the interests of the post-developmental state. In fact, post-developmental governance acknowledges and seeks to develop people's capacities to act and think in certain ways in order to utilise these capacities for broader social purposes.

The post-developmental state (when compared to Castells' (1996) developmental state) is characterised by greater flexibility and responsiveness to the rapidly changing economic conditions of 'fast capitalism.' It manages shifting relations between the state and society by managing populations through flexible new forms of governmentality (Foucault 1991) that place greater responsibility for governance on individuals. Imagination plays an important role in these adaptive, flexible forms of governance. If people can see themselves as lifelong learners, ready to innovate, change, and adapt for the good of the community or nation or in accordance with the shifting demands of capital, they are more likely to play the role of a shape-shifter, ever-ready to do whatever when called upon by their leaders. Imagination, in this case, goes both back to the past and into the future.

As Ong points out, 'Globalisation has induced new imagined communities that do not merely stress continuity, but also a resurgence of ancient traditions that go beyond past achievements to meet new challenges of modernity' (Ong 2000, 59). Imagination thus serves to connect people with an imagined past and disciplines them to participate in an imagined future characterised by great risk, change, and uncertainty.

In her more recent work, Ong draws on Deleuze to argue that this creates 'an environment of constant modulation shaped by latitudinal flows in which the human subject is in continuous training and monitored for persistent self-management' According to Ong, this leads to 'latitudinal citizenship ... shaped by this capacity to respond continually to the dynamism of the space of flows, to respond quickly and with agility to the ever changing conditions and requirements of market trajectories'(Ong 2006, 124). Latitudinal educational systems, likewise, might be expected to embody the capacity to respond continually, quickly, and with agility to changing conditions, requirements and trajectories. The call for innovation, responsiveness, diversification and a process orientation can be seen as moving Singapore's educational system toward this sort of agility.

Also, since meritocracy and hierarchy remain fundamental aspects of Asian tigers, such as Singapore, citizens are differently and differentially articulated to global capital and 'subjected to different kinds of surveillance and in practice enjoy different sets of civil, political, and economic rights' (Ong 1999, 215–16). Or, as Castells argues, global flows and state policy contain asymmetries which form structures of belonging and possibility (Castells 1999). This can be seen in terms of the different streams and choices available to students in the Singaporean system. As Gopinathan suggests,

> The cognitive elite, already the ones best placed to benefit from Singapore's streamed system of education, will gain further advantage from the integrated programme and the well-to-do will benefit from the establishment of private schools and universities, and opportunities to seek education abroad (Gopinathan 2007, 67).

The technical education stream prepares students for other types of labour power and articulates a different relationship with global capital. Different sectors of the population receive uneven distributions of services and care and 'are subject to different technologies of regulation and nurturance, and in the process assigned to different social fates' (Ong 2000, 57).

Post-developmental educational reform, then, aims to systematically provide pastoral care in which the state helps identities and communities develop capacities for flexibility, adaptability, responsiveness, and continual learning that will help them maintain flexible relationships with global capital and the nation state and manage uncertainty. It uses an array of strategies that help the state define and continually re-define its relationship with communities and identities in uncertain times. We now turn to the consequences of such flexible arrangements.

New Tensions and Challenges for the Post-developmental State

New tensions and challenges arise for the post-developmental state and post-developmental educational reform. Three in particular are described below: challenges to identity formation and citizenship education, tensions central to encouraging critical thinking and innovation, and potential challenges to the social discipline central to post-developmental governance.

Whither the Nation State in an Age of Multiple Flows and Allegiances?

New imperatives of globalisation have created pressures on nation states and educational systems to form allegiances and create identities amenable to national projects. As K. Mitchell argues, there has been both 'growing pressures for greater educational standardization and accountability ... [and] the creation of the tolerant, "multicultural self", a more individuated, mobile, and highly tracked ... "strategic cosmopolitan"' (Mitchell 2003, 387). This 'strategic cosmopolitan', however, is not motivated by ideals of national unity, but by 'understandings of global competitiveness, and the necessity to strategically adapt as an individual to rapidly shifting personal and national contexts ... i.e., [they are] individuals oriented to excel in ever transforming situations of global competition, either as workers, managers, or entrepreneurs' (Mitchell 2003, 387). In other words, the flexible identity required by globalisation is a different type of identity than that required by the developmental state and may be much more difficult for nation states to manage and hold in allegiance.

Increasingly, transnational problems and issues such as global warming and pollution, disease, war, terrorism, financial crisis, poverty, increasing inequality and oppression seem to require transnational solutions and forms of citizenship. Traditional communities and notions of national citizenship are being destabilised by forces of global capitalism, secularisation, new technologies and media, and democracy (Parker, Nonomiya and Cogan 1999). It remains to be seen to what extent flexible global cosmopolitan identities are viable in the nation state.

New 'technoscapes', 'ideoscapes' and 'mediascapes' (Appadurai 1996) make possible new imagined identities and communities. Transnational movements of ideas and media that are increasingly beyond governments' capacities to filter and control are powerful shapers of identities and pose the potential for transnational allegiances and communities. These ideas and media comprise an out-of-school curriculum and offer powerful content and pedagogies that vie with official curricula and pedagogies. Students have access to all kinds of technologies that are fundamentally social in nature and make transnational communication, expression, and mobilisations possible. Young people live increasingly digital lifestyles and have opportunities to develop identities and communities in gaming environments, through blogging, in online chat rooms, through popular Web sites like YouTube and through interaction in social networking Web sites such as Facebook and MySpace. They are able to create and share content that has the potential to challenge official discourses.

Emerging technologies and media thus pose new problems for nation states at the same time they are embraced. While new media and technology have become integral to the educational landscape, there are several negative aspects of digital content that nation states and schools alike must deal with: students may come upon misinformation (e.g., wrong or incomplete information); 'malinformation' (e.g., information that is harmful, such as pornography or bomb making); 'messed up' information (e.g., information that is badly presented, unorganised and unusable); and useless information (e.g., information that is of little relevance or use) (Burbules and Callister 2000). There are also 'counterideologies' that challenge official narratives and knowledge. For example, transnational human rights groups such as *Reporters without Borders* have challenged Singapore's crackdown on bloggers and cyberdissidents and defamation suits that challenge press freedoms (Reporters without Borders 2007). Gay rights and alternative lifestyle movements also challenge more traditional conceptions of identity and community. How Singapore's government will engage these oppositional or alternative identities and communities represents a core challenge to the post-developmental state.

What are the Limits of Critical Thinking, Openness and Innovation?

Aaron Koh challenges notions of critical thinking embodied in Singapore's curricular reforms. He asks to what extent critical thinking, openness, and innovation are possible in a system dominated by an examination culture and hierarchy. Also, critical thinking in Singapore has been cast as a process of technical problem solving requiring procedural skills such as analysing, interpreting, and evaluating information rather than a criticality that questions 'the power structures and the ideological constructions of truth and belief' (Koh 2004, 339). To what extent will Singaporean officials and educators allow a broader and more fundamental criticality that challenges official discourses and accepted assumptions and practices?

Catherine Lim explains how Singapore has managed the tension of advocating critical and creative thought and innovation while setting firm limits on what is acceptable. Commenting on the 'out-of-bound markers' that shape the limits of public discourse, she argues that they are purposefully left vague to promote a 'general sense of fear, hardly definable and therefore easily challenged by the Government' as non-existent (Lim 2006, 89). In this 'atmosphere of continuing anxiety, there will be continuing self-censorship. The greater the Government's efforts to increase material prosperity, the more irrelevant and even harmful will be seen the role of the political dissident' (Lim 2006, 91). The ideal and supposed practice of meritocracy also serves to winnow out talents that might challenge existing norms and assumptions. Those that rock the boat and go against accepted values and wisdom are unlikely to rise in the ranks, either in public or business endeavours. These limits may hinder critical and creative thought and innovation and discourage a creative class that is considered essential to economic development in post-industrial societies such as Singapore (Florida 2002; 2005).

A patron-client relationship (Scott 1976) is also invoked by government officials whereby the patron state provides economic security and collective safety in exchange for allegiance and compliance from its clients. Ong argues that crises have been used strategically to maintain this patron-client relationship: 'Observers have noted that the Singaporean state maintains power through orchestrating crises that become opportunities for the government to identify 'threats' to state security, to marginalise potentially dissenting groups, and to instill self-surveillance in a population induced to feel continually under siege' (Ong 1999, 72). This allows the government to provide what seems to matter most to people – safety, security, and prosperity – in exchange for economic discipline and social conformity.

Lim argues that two other tactics are used to pacify citizens and manage dissidence: support for freedom of expression in the arts and humanitarian concern for the underclass and underprivileged (Lim 2006). Dissidents are allowed to vent and freely express themselves artistically, although there are limits here as well as suggested by the Singapore Media Development Authority's denial of two artworks making up the third annual Singapore gay pride festival, 'IndigNation' (Kolesniko-Jessop 2007). The government's social policies to help the poor, such as the 'New Singapore Shares' in 2001, the 'Economic Restructuring Shares' in 2003, and the 'Workfare' scheme in 2006 (Neo and Chen 2007), have helped deflect criticism that the government was not doing enough to help the poor and underskilled adjust to economic restructuring.

The mantra of good governance touted by government officials and the media also supports the notion that governing should be left to the ruling elite. It minimises the need for an active civil society or the political engagement of citizens. Instead, people 'become so dependent on the Government for making decisions for us, for thinking for us, and so used to our comfortable lives, that any major change and adjustment will be viewed with alarm' (Lim 2006, 94). Indeed, perhaps Singapore's citizens are so well disciplined and trained that when anyone gets out of line or challenges the tight strictures of society, the people themselves will censure and rebuke offenders. The government can rely on its people to keep order and enforce out-of-bound markers.

To what extent politically sensitive issues such as income disparity, class stratification, or racial stratification based on these disparities can be critically analysed also remains open to question. The PAP's educational philosophy seems to be that intelligence is largely determined by genetic makeup and that the 'talented few' should be the natural leaders of Singapore (Rahim 1998). As Prime Minister Goh noted, 'we cannot narrow the [income] gap by preventing those who can fly from flying ... Nor can we teach everyone to fly, because most simply do not have the aptitude or ability' (Goh 1996). To what extent 'the class-divide between the well-educated, privileged, globally-mobile elite, on the one hand, and the working class majority, on the other' will be addressed remains to be seen (Tan 2004, 87). For example, ethnic Malay and Indian minorities form a disproportionately large percentage of lower income groups and a small percentage of the high-income strata (Tan 2004). These disparities are likely to increase with the asymmetrical development that accompanies globalisation (Stiglitz 2006).

Such tensions between the call for critical thought and innovation and limits set by 'out-of-bounds markers' are likely to be felt in educational settings. Whether issues typically considered sensitive such as income inequality and ethnic disparities are open to critical thinking and innovative ideas is one area that needs to be addressed. Limits to alternative perspectives that challenge fundamental assumptions and official discourses provide another problematic area in an age of flexibility.

Flexibility, Consumerism and the Erosion of Discipline

The flexible citizen is not only a flexible producer and flexible learner able to adapt quickly to the changing conditions of global capitalism, they are also flexible consumers ever ready to purchase the latest goods, services, and experiences. As Appadurai notes, 'consumption has become the civilising work of post-industrial society' (1996, 81) as consumers' imaginations and desires are disciplined by new commodities that make up their lifestyle choices. Appadurai argues that consumption has become a new form of labour and social discipline:

> The labor of reading ever-shifting fashion messages, the labor of debt servicing, the labor of learning how best to manage newly complex domestic finances and the labor of acquiring knowledge in the complexities of money management. This labor is ... directed at producing the conditions of consciousness in which *buying* can occur ... This inculcation of the pleasure of *ephemerality* is at the heart of the disciplining of the modern consumer (1996, 82–3).

Consumers must keep abreast of the latest changes, new trends, new technologies, and new services and experiences. Flexible education that creates the adaptive, flexible citizen also serves to create the flexible consumer. Everything is commodified and even education itself becomes a commodity rather than a public good. With the increased marketisation of education, parents and students seek advantages in an educational system that stratifies, ranks and sorts, and helps some gain competitive advantage. In particular, 'it is the elite parents that see the most to gain from the special distinctions offered by a stratified educational system, and therefore they are the ones who play the game of academic one-upmanship most aggressively' (Labaree 1997, 54). As a result, education is valued for extrinsic rewards, such as status, prestige, and reputation, rather than for intrinsic values. Labaree argues that this consumer conception of education has resulted in an overriding concern for 'contest mobility' in which winning is emphasised over learning and private gain over efficiency: 'The essence of schooling then becomes the accumulation of exchange values (grades, credits, and credentials) that can be cashed in for social status rather than the acquisition of use values (such as the knowledge of algebra or the ability to participate in democratic governance)' (Labaree 1997, 67). Similarly, league tables serve the purpose of having exchange value that can be used for comparative purposes so that 'clients' know how their 'product' stacks up against other 'products.' Data, such as that provided in the school league tables helps clients consider the quality of the education they are consuming and how it compares to (and prepares students for) other educational opportunities.

Singapore's reforms have tried to balance flexibility with discipline (Ministry of Education 2005). However, flexible consumerism may work to erode social discipline. In other words, consumerism may not instill the kind of social discipline that continues to provide competitive economic advantage. As the social historian, Christopher Lasch argues, mass consumption and rampant consumerism

> tend to discourage initiative and self-reliance and to promote dependence, passivity, and a spectatorial state of mind both at work and at play. Consumerism is only the other side of the degradation of work – the elimination of playfulness and craftsmanship from the process of production … The state of mind promoted by consumerism is better described as a state of uneasiness and chronic anxiety. The promotion of commodities depends, like modern mass production, on discouraging the individual from reliance on his own resources and judgment: in this case, his judgment of what he needs in order to be healthy and happy. The individual finds himself always under observation, if not by foremen and superintendents, by market researchers and pollsters who tell him what others prefer and what he too must therefore prefer, or by doctors and psychiatrists who examine him for symptoms of disease that might escape an untrained eye (1984, 27–8).

When combined with the forfeiture of public decision making to the ruling elite who manage the corporate state, such consumerism may interfere with Singaporeans capacities to innovate, create, and think critically. Or it may lead to unease and anxiety that is directed at the ruling class. It may result in a lessening of the social discipline desired by the post-developmental state.

Conclusion

Whether or not TSLN reforms such as National Education and the implementation of the new Social Studies curriculum 'are more about attempts by governing elites to maintain power in increasingly challenged contexts (by forces such as globalisation), than a genuine concern for better educating young people' (Sim and Print 2005, 65) is a lingering question. Official rhetoric has called for greater critical and creative thinking, innovation and openness. However, as Tan and Gopinathan have pointed out,

The larger problem for Singapore's educational reform initiative is that Singapore's nation-building history resulted in an omnipresent state that cherishes stability and order. A desire for true innovation, creativity, experimentation, and multiple opportunities in education cannot be realized until the state allows civil society to flourish and avoids politicizing dissent (2000, 10).

Will continuing limits on dissent and civil society act as a brake on economic growth and prosperity in the new economy?

Charles Taylor describes the neo-liberal view of economic life that accompanies globalisation as a modern social imaginary that 'seems the only possible one, the only one that makes sense' (Taylor 2004, 17). It's hard to imagine alternatives to neo-liberal perspectives and post-developmental governance in an age of accelerating globalisation. Widespread transformations, such as those accompanying globalisation, shape how people think of themselves, their communities, and their experiences. Educational systems are caught up in these transformations and seek to cultivate the capacities that are necessary for people to participate in such transformations. These transformations also entail identity shifts. As Taylor points out, however, transformations usually distance or 'disembed' people from their traditions and communities. As a result, new efforts are made to help people develop a sense of belonging.

Or as Zygmunt Bauman argues, in the age of globalisation localities lose 'their meaning-generating and meaning-negotiating capacity and are increasingly dependent on sense-giving and interpreting actions which they do not control' (Bauman 1998, 3). However, Bauman offers a way to manage these disruptions through a critical questioning of 'the ostensibly unquestionable premises of our way of life' (Baumann 1998, 5). Perhaps such questioning can restore people's meaning-generating and meaning-negotiating capacities. Singaporeans pride themselves on being pragmatic, but there are other forms of pragmatism besides the sort of technical, instrumental pragmatism valued in Singapore. Pragmatism also calls for practices that resist closure and social rigidity. R.J. Bernstein notes that the heart of pragmatism is a 'willingness to talk, to listen to other people, to weigh the consequences of our actions upon other people' (Bernstein 1991, 198). This sort of openness and inquiry creates a set of flexible and growing habits that allows for a more rigorous examination and determination of possible courses of action that will benefit a greater number of interests. Such pragmatism supports experimentation and the testing of hypotheses; it welcomes diverse perspectives and is not afraid of opposing views; it is based on the values of freedom, mutuality, and respect and continually critiques claims to finality or closure. This sort of pragmatism invites everyone to participate in imagining and deciding their collective futures. It is based on the sort of critical questioning advocated by Bauman. A rigorous questioning of the premises of globalisation and post-developmental educational reform is urgently needed.

References

Agger, B. (1989), *Fast Capitalism: A Critical Theory of Significance* (Urbana, IL: University of Illinois Press).

Agger, B. (2004), *Speeding Up Fast Capitalism: Culture, Jobs, Families, Schools, Bodies* (Boulder, CO: Paradigm).

Appadurai, A. (1996), *Modernity at Large: Cultural Dimensions of Globalization* (Minneapolis: University of Minnesota Press).

Bauman, Z. (1998), *Globalization: The Human Consequences* (New York: Columbia University Press).

Bernstein, R.J. (1991), *Beyond Objectivism and Relativism: Science, Hermeneutics, and Praxis* (Philadelphia: University of Pennsylvania Press).

Bourdieu, P. (1977), *Outline of a Theory of Practice* (Cambridge: Cambridge University Press).

Brown, P. and Lauder, H. (2001), *Capitalism and Social Progress: The Future of Society in a Global Economy* (Basingstoke: Palgrave).

Burbules, N.C. and Callister, T.A. Jr. (2000), *Watch IT: The Promises and Risks of New Information Technologies for Education* (Boulder, CO: Westview Press).

Burchell, G. et al. (eds) (1991), *The Foucault Effect: Studies in Governmentality* (Chicago: University of Chicago Press).

Castells, M. (1996), *The Rise of the Network Society* (Oxford: Blackwell Publishers).

Castells, M. (1999), 'Flows, Networks, and Identities: A Critical Theory of the Informational Society' in Manual Castells et al. (eds).

Castells, M. et al. (eds) (1999), *Critical Education in the New Information Age* (Lanham, MA: Rowman and Littlefield Publishers, Inc.).

Cope, B. and Kalantzis, M. (eds) (2000), *Multiliteracies: Literacy Learning and the Design of Social Futures* (London: Routledge).

Eng, L.A. (ed.) (2004), *Beyond Rituals and Riots: Ethnic Pluralism and Social Cohesion in Singapore* (Singapore: Eastern Universities Press).

Eng, L.A. (ed.) (2006), *Singapore Perspectives 2006; Going Glocal: Being Singaporean in a Globalised World* (Singapore: Institute of Policy Studies).

Florida, R. (2002), *The Rise of the Creative Class. And How It's Transforming Work, Leisure and Everyday Life* (New York: Basic Books).

Florida, R. (2005), *The Flight of the Creative Class. The New Global Competition for Talent* (New York: HarperCollins).

Foucault M. (1991), 'Governmentality', in Graham Burchell et al. (eds).

Gee, J.P. (2000), 'New People in New Worlds: Networks, the New Capitalism and Schools', in Cope and Kalantzis (eds).

Gopinathan, S. (1994), 'Educational Development in a Strong State: The Singapore Experience', *Australian Association for Research in Education Conference*, Newcastle, Australia.

Gopinathan, S. (2007), 'Globalization, the Singapore Developmental State and Education Policy: A Thesis Revisited', *Globalization, Societies and Education* 5:1, 53–70.

Green, A. (1997), 'Education and State Formation in Europe and Asia', in Kennedy (ed.).

Holmes, D.R. (2000), *Integral Europe: Fast-capitalism, Multiculturalism, Neofascism* (Princeton, NJ: Princeton University Press).

Kennedy, K.J. (ed.) (1997), *Citizenship Education and the Modern State* (London: Falmer Press).

Khong, L.Y.L. (2004), 'School-Stakeholder Partnerships: Building Links for Better Learning', in J. Tan and P.T. Ng (eds).

Koh, A. (2002), 'Toward a Critical Pedagogy: Creating "Thinking Schools" in Singapore', *Journal of Curriculum Studies* 34:3, 255–64.

Koh, A. (2004), 'Singapore Education in "New Times": Global/local Imperatives', *Discourse: Studies in the Cultural Politics of Education* 25:3, 335–49.

Labaree, D. (1997), 'Public Goods, Private Goods: The American Struggle over Educational Goals', *American Educational Research Journal* 34:1, 39–81.

Lasch, C. (1984), *The Minimal Self: Psychic Survival in Troubled Times* (New York: Norton).

Lave, J. (1996), 'Teaching, as Learning, in Practice', *Mind, Culture and Activity* 3:3, 149–64.

Lim, C. (2006), 'Managing Political Dissent: Uniquely Singapore', in Eng (ed.).

Mitchell, K. (2003), 'Educating the National Citizen in Neoliberal Times: From the Multicultural Self to the Strategic Cosmopolitan', *Transactions of the Institute of British Geographers* 28:4, 387–403.

Neo, B.H. and Chen, G. (2007), *Dynamic Governance: Embedding Culture, Capabilities and Change in Singapore* (Singapore: World Scientific Publishing).

Ng, P.T. (2005), 'Introduction', in J. Tan and P.T. Ng (eds).

Ng, P.T. (2005), 'Innovation and Enterprise', in J. Tan and P.T. Ng (eds).

Ong, A. (1999), *Flexible Citizenship: The Cultural Logics of Transnationality* (Durham: Duke University Press).

Ong, A. (2000), 'Graduated Sovereignty in Southeast Asia', *Theory, Culture, and Society* 17:4, 55–75.

Ong, A. (2006), *Neoliberalism as Exception: Mutations in Citizenship and Sovereignty* (Durham: Duke University Press).

Parker, W., Nonomiya, A. and Cogan, J. (1999), 'Educating World Citizens: Toward Multinational Curriculum Development', *American Educational Research Journal* 36:2, 117–46.

Quah, J.S.T. (2000), 'Globalization and Singapore's Search for Nationhood', in L. Suryadinata (ed.).

Rahim, L.Z. (1998), *The Singapore Dilemma: The Political and Educational Marginality of the Malay Community* (New York: Oxford University Press).

Scott, J.C. (1976), *The Moral Economy of the Peasant* (New Haven, CT: Yale University Press).

Sim, J.B.-Y. and Print, M. (2005), 'Citizenship Education and Social Studies in Singapore: A National Agenda', *International Journal of Citizenship and Teacher Education* 1:1, 58–73.

Stiglitz, J.E. (2006), *Making Globalization Work* (New York: W.W. Norton).

Suryadinata, L. (ed.) (2000), *Nationalism and Globalization: East and West* (Singapore: Institute of Southeast Asian Studies).

Tan, E. (2004), '"We, the Citizens of Singapore …": Multiethnicity, its Evolution and its Aberations', in Eng (ed.).

Tan, J. and Gopinathan, S. (2000), 'Education Reform in Singapore: Towards Greater Creativity and Innovation?', *NIRA Review* 7:3, 5–10.

Tan, J. and Ng, P.T. (eds) (2005), *Shaping Singapore's Future: Thinking Schools, Learning Nation* (Singapore: Prentice Hall).

Taylor, C. (2004), *Modern Social Imaginaries* (Durham and London: Duke University Press).

Internet-based References

Daipi, H. 'Keynote Address "Achieving a Social and Moral Balance in Globalization" at ITE College East', 31 August 2005, *Ministry of Education*, <http://www.moe.gov.sg/speeches/2005/sp20050831.htm>.

Goh, C.T. 'Prepare Our Children For the New Century: Teach Them Well', Teachers' Day Rally, 8 September 1996, *Contact, Ministry of Education*, <http://www. moe.gov.sg/corporate/contactonline/pre-2005/rally/speech.html>.

Goh, C.T. 'Shaping our Future: Thinking Schools, Learning National Nation', 2 June 1997, *Ministry of Education* <http://www1.moe.edu.sg/Speeches/020697. htm>.

Kolesnikov-Jessop, S. 'Singapore Gays Allowed a Step Forward, but Pushed Two Back', *International Herald Tribune*, [website], 2 October 2007, <http://www. iht.com/articles/2007/08/02/news/singapore.php>.

Lee, H.L. 'The Launch of National Education: "National Education"', 17 May 1997, *Ministry of Education* <http://www.moe.gov.sg/speeches/1997/170597. htm>.

Luke, A. 'Curriculum, Ethics, Metanarrative: Teaching and Learning Beyond the National', *Curriculum Perspectives*, 2002, <http://wwwfp.education.tas.gov. au/English/luke.htm>.

Ministry of Education, Singapore, <http://www.moe.gov.sg/>.

Ministry of Education, *The Next Chapter: Innovation and Enterprise, Taking I and E Forward*, 2005, <http://www.moe.gov.sg/bluesky/The_Next_Chapter.pdf>.

Reporters Without Borders, 'Singapore: Annual Report, 2007', 2007, <http:// www.rsf.org/article.php3?id_article=20796>.

Teo, C.H. 'Launch of the Masterplan for IT in Education', 28 April 1997, <http://www.moe.gov.sg/edumall/mpite/mpite_launch/speech0.htm>.

Teo, C.H. '"World Singapore" – Singapore in the World; The World in Singapore', 19 April 1997, <http://app.sprinter.gov.sg/data/pr/20070419990.htm>.

Tharman, S. 'Leading Schools in a Broad-Based Education', 10 April 2004, *Ministry of Education*, <http://www.moe.gov.sg/speeches/2004/sp20040410a.htm>.

Wong, A. 'Address at the Opening Ceremony of the Primary Social Studies Symposium at York Hotel, Carlton Hall', 13 March 2000, *Ministry of Education*, <http://www.moe.gov.sg/speeches/2000/sp13032000a.htm>.

Chapter 5

Morphogenesis and Hybridity of Southeast Asian Coastal Cities

Johannes Widodo

Morphogenesis of Southeast Asian Cosmopolitan Cities

Within Southeast Asia, cosmopolitan settlements have been growing and developing around the coastal areas of the South China Sea, Java Sea, and Malacca Strait (the 'Mediterranean of Asia', see Figure 5.1), since the beginning of the inter-insular and inter-continental trades dating at least from the first century. In the past, this region was politically unified under various maritime kingdoms, continuously shaped and enriched by various cultural layers and elements, constantly nurtured and developed throughout its history.

From the fertile Mekong delta region, the source of rice culture of Southeast Asia, the first kingdom of Funan was established around the Mekong delta around 100–600 CE, followed by several other inland kingdoms of Chenla (600–790 CE), Pagan (849–1287 CE), Khmer (790–1431 CE), Ayuthaya (1350–1767 CE), and Champa (192–1471 CE). Srivijaya maritime power (600–1290 CE) took effective control over the main trading routes of Melaka strait and Java Sea. It was the period when Hindu–Buddhist cosmology and its materialisation of culture spread across the region. Islam entered and spread throughout Southeast Asia through various trading routes. From the west the Arabs, Persians, and Indians (Gujarat and Tamil) came to Malay Peninsula and the west coast of Sumatra, to northern coast Java and all over the archipelago in the thirteenth to fourteenth centuries. From the north came the Chinese Muslim traders, peaking especially during the voyage of Ming dynasty's admiral Zheng He (1405–1433) in the fifteenth century. The Chinese traders and immigrants from Southern China had been settled down in the coastal cities of Southeast Asia since as early as the twelfth century, and getting more intensive from the fifteenth century onward.

Located thus, at the crossroads of global maritime trading routes, Southeast Asia has been very open towards the various cultural influxes. These cultures were then transplanted, adopted, absorbed and nurtured locally, then expressed into unique but yet closely linked culture, language, architecture, and artifacts. The settlements are formed by complex layers of various cultures, ideologies, economies, and ecosystems, and manifested in the hybrid urban morphology and architectural typologies. Here cultural and geographical 'boundary' is always blurring, overlapping, or intersecting, and has never been clearly defined.

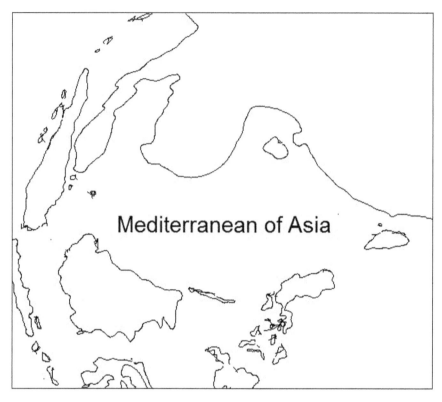

Figure 5.1   Mediterranean of Asia
Source: Figure by author, Johannes Widodo.

At the dawn of Southeast Asian urban maritime civilisation, the vessels from China, Japan, and Ryukyu[1] sailing to the south made use of the northern monsoon between January and February, returning home by the southern monsoon between June and August. Indian and Arab ships went eastward by the southwest monsoon between April and August, and returned by the northeast monsoon from December. During the cyclone periods or the changing monsoon seasons, these traders stayed in Southeast Asian ports (such as Samudra Pasai and Melaka) and inhabited the markets, while waiting for their trading partners from the other parts of the world. Metaphorically the city is like a boat or a vessel, loaded by people, goods, activities, rituals, and symbolism. The trading ships and immigrant boats were not only carrying people and goods, but also conveying cosmological and

---

1   The kingdom of Ryukyu, located at the currently known Okinawa islands, appeared around the fifteenth to sixteenth century growing as intermediary trading hub between East Asia (China and Japan) with Southeast Asia, before the European disrupted and dominated the maritime trade in this region.

geometrical patterns (like architectural typologies, urban grids, and hierarchy of spaces) from original sources into the new landscapes, implanting new layers in the new lands. People from different places, islands or continents are kept moving, communicating, and intermingling, influencing each other and producing hybrid, fused, diverse architecture and material culture.

For more than two millenniums of urban history, cities in this region have been demonstrating their ability to preserve primary elements and basic morphological patterns. Never ending processes of layering, transformations, and hybridisation, is probably the best way to describe its dynamic and sustainable characters. Diversity, eclecticism, fusion, acculturation, adaptation, can perhaps describe the nature of Southeast Asian architecture and urbanism from the past into the present and hence into the future.

Layering Process of Urban Morphology

The vernacular building tradition in Southeast Asia is the outcome of local climate, building materials and techniques, also indigenous beliefs and rituals. The people who live inland are mostly involved in agriculture, with rice-culture as the most dominant activity. Those who live in coastal areas are traders and seafarers. The architecture in this region is the reaction to equatorial and tropical warm-humid climates. Architecture is also adapted to earthquakes, especially within an archipelago which is continuously rocked by active volcanoes and continental plate movements. The building construction method is similar to the shipbuilding technique, and the flexible structural system is well adapted to absorb destructive forces of the earthquake.

Rivers systems have been the lifeline of human settlement. Early settlement that bore the seeds of urbanity appeared near the waterfront, as the connection point between the outside worlds and the interior hinterland. The waterfront settlement nucleus in the Malayan, Indonesian, and Cambodian contexts is called *Kampung*. According to some locals, *Kampung* (in languages currently described as 'Bahasa Malaysia' and 'Bahasa Indonesia'), or *Kompong* (in Cambodian) originally refers to the area on the riverbank near the landing point and on the path to the settlement further uphill from the waterfront.

In subsequent centuries, Indian cosmology (Hinduism and also Buddhism) was transmitted from India to Southeast Asia and East Asia through trades and migrations, applying a new layer of ordering principles and meaning into vernacular spaces and structures. It is believed that human lives exist in between vertical and horizontal universal orders, metaphorically summarised in the tripartite hierarchical sub-division of upper-middle-lower (see Figure 5.2). This cosmological sub-division follows the metaphor of the human body, the head, the torso, and the feet, parallel to the metaphor of the universe, sky, ground, and underworld. This is known as the concept and ordering principles of the 'Mandala'.[2]

---

2   'Mandala' is a Sanskrit term meaning 'circle' or 'completion'.

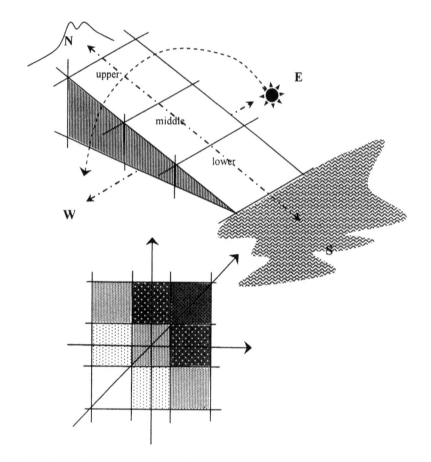

Figure 5.2   Superimposition of two tripartite cosmological hierarchies of space

Source: Sketch by author, Johannes Widodo.

The physical remains of the cultural heritages from this period were almost vanished due to the non-durable materials such as brick and timber they used in the building construction, but many Hindu–Buddhist made of stones temples still survived (such as Angkor in Cambodia, Borobudur and Prambanan in Java). However from the archeological findings it is evident that the layout of the cities in this region showed clear underlying Indian cosmological principles, harmoniously merged with the local vernacular planning and design traditions.

Islam entered and spread throughout Southeast Asia through various trading routes in the thirteenth to fourteenth centuries. The Arab, Yemeni, Gujarati, and Tamil traders from West Asia and South Asia were the earliest agents of the spread of Islam in Southeast Asia, followed by the Chinese Muslims, the latter especially during the voyages to Southeast Asia and the Indian Ocean of Admiral Zheng He of the Ming dynasty in the fifteenth century. The process began from the urban centres along the northern coast of Java (Gresik, Tuban, Demak, Semarang) and northern tip of Sumatra (Samudra Pasai in Aceh), then spread to other coastal port cities in the Malay peninsula and Indonesian archipelago (Melaka, Palembang, Makassar, Banjarmasin, Ternate, Tidore, Ambon, etc.), and southern Philippines (Mindanao, Palawan, and the islands around Sulu sea).

Islam introduced new typology and vocabulary in Southeast Asian architecture and urban forms such as the Islamic orientation towards the Qiblat (praying orientation towards Mecca). The natural and peaceful fusion processes of the new Islamic design principles into the building and urban spatial typology from the previous periods took place during this period. The hybrid architectural style is a tangible manifestation of the cosmopolitan spirit and the tolerant nature of the Southeast Asian urban culture in embracing and incorporating new foreign elements. Patrons, artists and builders from different ethnic and cultural groups worked together and coordinated their artistry and skill to construct new and unique building tradition and architectural form, for example Kampung Kling mosque in Melaka (Figure 5.2).

Figure 5.3   Kampung Kling mosque in Melaka

Notes: Tangible architectural manifestation of cultural fusion and hybridity of Javanese, Indian, Malay, Chinese, Portuguese, Dutch and other Architectural elements and detailing.

The coastal region of Southeast Asia was the first place where new cities and coastal settlements appeared as a result of this international trading network. During the cyclone periods or the changing monsoon seasons, the traders stayed in Southeast Asian ports, while waiting for their trading partners from the other parts of the world to come. During their stay the crew and passengers of the ships populated the city and mingled with the local population. Many of the early Chinese colonies were developed near river estuaries closed to the pre-existed indigenous villages. Some of these early settlements then grew into flourishing entrepôts[3] (such as Pattani, Melaka, Palembang, and Semarang) thanks to the vibrant international maritime trading. The Chinese architectural elements blended with the local-vernacular design features created numerous variations of fusion building styles.

The voyages of Admiral Zheng He to Southeast Asia and the Indian Ocean during the Ming Dynasty left tangible traces along the coastal regions of Southeast Asia in the form of Chinese settlements. These new coastal towns were well integrated into the pre-existing structure of the original coastal villages, mostly situated near the river estuaries and shorelines. The morphology of these cosmopolitan settlements consisted of an interrelated double nuclei. It is a common occurrence in coastal Southeast Asia that an old Chinese temple is located adjacent to an ancient mosque within the old urban core, close to the waterfront, at the middle of a multi-racial cosmopolitan community. Both of these areas were separated but interconnected by a market place not far from the harbour. The market place was the common urban centre, a meeting place for the locals and foreigners to meet and to exchange. It was a public place with a strong cosmopolitan character. Unique identity, belief, and material culture of each group were preserved and nurtured, and at the same time a new communal hybrid identity would be created and developed, based on mutual respect and the spirit of tolerance.

On every Southern Chinese immigrant boat, a special shrine for Mazu[4] was installed to safeguard the compass, the steering wheel, the sailing direction, and all passengers aboard. Once the boat reached the destination in the South Seas and they decided to settle down, the ship would be dismantled and the shrine would be reconstructed near the landing place at the shore or riverfront. The new settlement's spatial structure of the diasporic southern coastal Chinese was a reconstruction of the cosmological pattern of the immigrant boat, where the Mazu temple was placed at the end of the axis facing the harbor and two masts were placed in front of the temple (such examples in Palembang, Melaka, Yangon, etc.). The memory embedded in the spatial-cosmological concept that was carried through the boat

---

3   'Entrepôt' is from the Latin words *inter* (in-between) and *postitum* (location), meaning a place located in-between. An entrepôt is a small town without or with very little local commodity. Almost all goods to be exchanged there were international goods in transit.

4   Mazu is protector goddess of fishermen, seafarers and immigrants from Southern China. Mazu worship spread to Southeast Asia through the waves of immigrations, and amplified by Zheng He voyages in the fifteenth century and after.

design, was preserved and transplanted into the new adopted land to form a new pattern urban nucleus. A small community of Chinese fishermen and traders then lived around this nucleus. They lived side by side with the local population and other communities, producing hybrid urban culture and architecture.

As an expression of respect to the local or indigenous spirits who guarded the place, the newly arrived Chinese immigrants paid their tribute to local objects such as a sacred stone, an old tree, or other object. Sometime later these objects were personified in the statue of a hybrid-deity, known as 'Datuk-Kong' – a peculiar combination of the Malay honorific title 'Datuk' and the Chinese 'Kong'.[5] The Datuk-Kong's shrine is regarded as the guardian of place, and is usually located at the thresholds or entrances to the town, neighbourhoods, temples and houses (Figure 5.4).

Figure 5.4   Datuk shrines

Notes: Datuk shrines (under the tree) guarding the Mazu Temple and Hainanese Association hall in Kuching (East-Malaysia). The two Datuks picture are inserted on the left.

---

5   'Datuk' is manifested as a Muslim indigenous old man sitting in a throne holding a walking stick or a traditional weapon: the 'Keris'. He has different names in different places, such as Datuk Awang, Datuk Haji, Kyai, or other local names across coastal regions of the Malay Peninsula, Kalimantan, Jawa, and Sumatra.

Besides these coastal settlements and cities in different parts of the Indonesian archipelago, where the majority of Chinese diaspora settlers hailed from the coastal regions of Southern China (Fujian, Guangdong, and Hainan), there is also the rather different settlement morphology of the Hakkas. The Hakka immigrants came from the mountainous region of Southern China (Guangdong and Fujian) during the colonial period in Southeast Asia, and many of them settled down in the mining towns in the Malay Peninsula, Kalimantan, Bangka and Belitung, especially after the failure of the Taiping revolution against the Manchu Dynasty in China.

Unlike their coastal counterparts, the Hakkas venerate gods of the mountain and land (Dabogong and other ancestral spirits) and are keen worshippers of Guandi (the legendary god of war manifested from the Tales of the Three Kingdoms), and also the Buddhist Goddess of Mercy, Kuanyin. Their settlements were built around these temples, which have another purpose as the community hall or ancestral hall. The settlement structure is based on a radial-concentric pattern, three-pronged main roads or axis pattern with the temple in the middle, guarded by a joss-house at every entrance to the settlement (Figure 5.5). This applied the similar

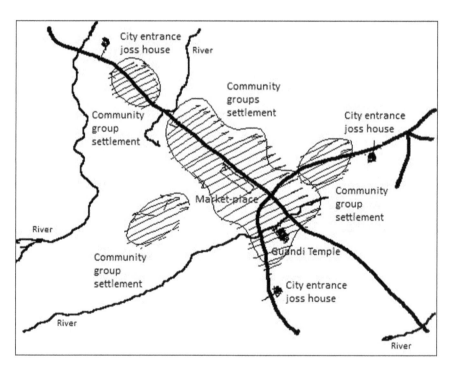

Figure 5.5   Diagrammatic map of Hakka mining town of Monterado (West Kalimantan) with a temple for Guandi at the city centre

Notes: The morphological pattern of this town follows the 'Tulou' planning principles.

Source: Map by author, Johannes Widodo.

spatial conceptual pattern of the original Hakka round multi-stories communal house 'Tulou' in Southern China, but manifested differently in the new contexts in Southeast Asia. The original basic concept is maintained and remembered, but the realisation is freely deconstructed, reinterpreted, reconstructed and localised.

Other popular-vernacular deities subsequently appeared, followed by the Taoist and Buddhist. The latter colonised or converted the pre-existed temples into Taoist or Buddhist temples, or they built new ones in the later stages of the urban development. They were followed by the Confucians and other religious sects, continuing hybridisation and layering processes within the multiracial settlements, adding into the assortment of religious temples, by mixing of various elements from different cultures and producing even richer blend of cultural materialisation. One good example of this acculturation process is the typical trader's house in Palembang, South Sumatra. The house plan and some of its construction detailing were of the southern Chinese courthouse origin, but the roof typology, open veranda, timber material, and its raised floor were taken from the local-vernacular vocabulary. Strong elements of Arab, Indian, and even European origins were further blended into this vernacular-cosmopolitan fusion dwelling typology, and became uniquely local (Figure 5.6).

Figure 5.6   House of the Chinese Captain in Palembang

Notes: The plan follows Chinese courtyard house pattern, the roof and main structures are of local Palembang vernacular style, the façade has Dutch construction elements, and the details are mixed of different building traditions.

Modernisation and Fragmentation

The European colonial powers had been expanding their hegemonic ambition since the late fifteenth century, initiated by the Portuguese (to India, Melaka, Java, eastern Indonesia, Philippines, Taiwan, Japan), followed by the Dutch (to India, Melaka, Indonesia, Taiwan, Japan), British (to India, Malaya peninsula, Bengkulu, Java, China), Spanish (Philippines), French (to Indochina, China), and Germans (to China). They introduced new building typologies, modern urban infrastructures, and new urban patterns, such as boulevards, streetscapes, façade, building techniques, and new functions (military establishments, public buildings, churches, urban squares and plazas, markets, railroads, stations, plantation houses, and many more).

At first, the four-season European design was directly transplanted and applied into tropical Southeast Asia with minor modifications. But then more responsive design solutions were invented, to adapt building and urban design into local warm-humid tropical climate, available local materials and technology, and contextual social-cultural conditions. European style buildings with deep verandah and ventilation holes, mixed with Chinese, Indian, Malay, Arab, and other design features, were added into the variety of local typologies and evolving into unique regional styles. As in the previous process, the European influences were easily and openly accepted into the vocabulary of Southeast Asian architecture and urbanism. The locals were also learning new methods and adopting new styles into their own, departing from their traditional rules, and embracing the modernisation process. The hybrid and varied nature of Asian modernity is thus manifested in the myriad architectural forms and styles, as a reflection of the actual ideological, political, and cultural realities.[6]

At the urban level, the ethnical segregation policy of dwelling areas according to different races was implemented in almost all colonial cities. The urban population was racially segregated, generally divided into three classes: the European, the 'oriental' (Chinese, Arabs, and other Asians), and the native population groups, allocated to separate settlement areas. Generally there was no clear demarcation to separate the racial zones, although in some cases there were rivers, walls, or

---

6 For example: the 'Euro-Vernacular' type which was adapted and developed in various parts of Asia as response to local climates and lifestyles which are different from the European context (the tropical British colonial bungalows in India and the Straits Settlements in peninsular Malaya, Dutch colonial villas and plantation houses in Indonesia), the 'Imperialist'-style which was forced and transplanted with no or little modification by the European (Portuguese, Spanish, Dutch, British, French, German, Russian in Asia) and also Asian colonial powers (Japanese in China and Korea), the 'Orientalist' style buildings (such as mosques built by the Public Works Department in Dutch-Indonesia, public buildings in French Indo-China, and colonial buildings in British Malaya), and the 'Pseudo-Western' type imitated or superficially copied from Western architectural elements by local architects.

roads, which functioned as physical boundaries (Figure 5.7). Nonetheless the non-physical legal boundaries caused an internal densification process within each restricted zone, especially apparent in the Chinese quarters, and the fragmentation of the originally coherent cosmopolitan city. The over-densification process would lead into environmental disasters and degradation of public health conditions for the whole city. The deterioration of the urban environment had forced the Colonial municipality governments to abandon their racial segregation policy and to launch infrastructure improvement programs in urban sanitation and utilities for the benefit of all community groups, such as the implementation of new building regulations

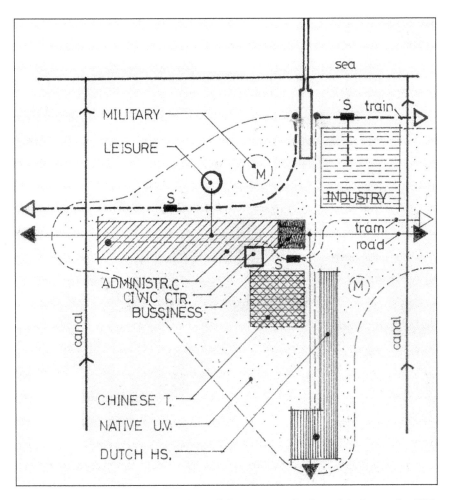

Figure 5.7   Morphologic model of Semarang (Indonesia) in early 20th century

Source: Diagram by author, Johannes Widodo.

and urban planning in the Straits Settlements (Melaka, Penang, Singapore) by the British, the Kampong Verbetering (improvement program) and the urban renewal plans in some Indonesian cities (for example in Batavia, Semarang, Surabaya) by the Dutch.

Since the beginning of the twentieth century, a rapid modernisation process has been taking place, triggered by the revolution in transportation and urban infrastructure, in order to cope with the rapid population, commerce, and industrial growth. The transportation revolution, started by the construction of railways followed by the introduction of automobiles and the construction of wider roads opened all isolations and broke up physical segregations. In a very short period of time the technological revolutions in many aspect of daily life such as electricity, gas, telephone and telegraph, newspapers, banks, post office, public transportation, had been transforming the colonial towns into modern and multi-functional cities. The rationality of function and the logic of economy have gradually replaced the politically and culturally motivated racial policy in the planning and design of Southeast Asian cities (see Figure 5.8).

Harbours were enlarged or upgraded, some industrial estates were developed, and the central business districts in the city centres were rejuvenated. Trade fairs were organised to accommodate the rapid growth in local and international

Figure 5.8   **Sultan Mosque in Labuhan Deli, Northern Sumatra, in hybrid Art-Deco style (1854)**

commerce and trade. Many fresh ideas from modern urban planners and architects were manifested into city plans, urban designs, and architectural styles, blended with the elements from the local, natural and cultural contexts. The port cities in Southeast Asia had grown up to the level similar to other modern port cities in the world of that period. Some cities across Southeast Asia such as Penang, Singapore, Medan, Batavia, Semarang, Surabaya, Makassar, and Manila were linked to the modern international maritime trade networks and developed into the major regional growth and distribution centers.

Hopes and idealism were put into the transformation and the future of the cities in Southeast Asia, but the global economic recession and the two World Wars which raged in the first half of twentieth century terminated these dreams. New political movements for independence and urban riots against the colonial powers proliferated all over the region. The economic recession and social-political instability had stopped some urban development projects and had left many parts of the city in deteriorated and dilapidated condition. The Second World War and the invasions of the Japanese Imperial army into East Asia and Southeast Asia gave the final blow to end the colonialism history in Southeast Asia, and changed the course of urban history and morphologyof this region. Almost 300 years of European colonisation in Southeast Asia had to come to an end. A new chapter of the Southeast Asian urban history would soon begin to emerge, riding the waves of decolonisation and the capturing the spirit of national independence.

Restoring and Reinventing the Cosmopolitan Spirit of Southeast Asian Cities

Southeast Asia is a dynamic source of identities, a place where great civilisations met, collided, and blended. It is like a complex web made of different elements and colours, but yet integrated as a continuous and coherent fabric. This richness and complexity is formed in a historical continuum from past to present.

Unfortunately, the current rapid economic growth has accelerated the cultural and physical transformation process, often leading to the fragmentation and destruction of old urban fabrics and resulting in the loss of identity and cultural amnesia. The layers of urban history and shared heritages which kept the shared memory of the whole community for many generations and centuries have been forgotten, and in many cases these have been erased completely, to be replaced with new alien pathological forms and functions.

The architecture of our cosmopolitan cities has many layers, morphological, sociological, and symbolical (form, function, and meaning). It is in the intersection of many disciplines from art to science, from philosophy to engineering. Therefore an inter-disciplinary approach is necessary in dealing with the current issues and problems. To ensure continuity and sustainability we need to empower the ordinary members of the community through training and education. The vanishing or dying craftsmanship and traditional skills should be revived and revitalised, and these have to be updated with the current technological advancement and contemporary innovations. Designers, academics, scholars, trainers, professionals have special roles as facilitators in community education through various ways. They should help the community to respect and to be sensitive to their own legacy, to teach them in understanding the existing ordinary fabrics and artifacts, to train them in discovering their own heritage, and to maintain the tradition and to promote the contemporary appropriation of their own inheritance. In this sense, conservation has special meaning and role in nurturing the community cultural continuum for the present and future generations, for the sake of restoring and reinventing the cosmopolitan spirit of our cities and citizens.

In this age of rapid changes and drastic transformations, there are some promising developments and hopes in the preservation, conservation and revitalisation of Southeast Asian urban and cultural heritages. In different cities across Asia many community based heritage movements, government initiatives, and private sector involvements in preserving, conserving, restoring, adapting, and revitalising the shared heritages are proliferating on unprecedented levels. Cities like Hoi An in Vietnam, Vigan in the Philippines, Luang Prabang in Laos, are given World Heritage city status because of their successful efforts to maintain their cultural and urban heritages.

New discourses of Southeast Asian architecture and urban studies have been flourishing across the region among the architectural schools and profession, and within the communities. With its long tradition in cultural diversity and inclusive creativity, reinvention and restoration of Southeast Asian memory, identity, continuity, and prosperity are well assured – along with the rapid changes and often unpredictable challenges.

References

Alisyahbana, S.T. (1987), *Bumantara – The Integration of Southeast Asia and its Perspectives in the Future* (Jakarta: Center of Southeast Asian or Bumantara Studies, Universitas National).

Engelhardt, R. (ed.) (2007), *Asia Conserved – Lesson Learned from the UNESCO Asia-Pacific Heritage Awards for Culture Heritage Conservation (2000–2004)* (Bangkok: UNESCO).

Ma, H. (1997 [1433]), *Ying-yai Sheng-lan: The Overall Survey of the Ocean's Shores* (Bangkok: White Lotus Press).

Rossi, A. (1984), *Architecture of the City* (Cambridge: MIT Press).

Widodo, J. (2002), 'Southeast Asia – Architecture', in Levinson, D., Christensen, D.K. (eds), *Encyclopedia of Modern Asia* (New York: Charles Scribner's Sons).

Widodo, J. (2004), *The Boat and The City – Chinese Diaspora and the Architecture of Southeast Asian Coastal Cities* (Singapore: Marshall Cavendish Academics).

Chapter 6

Nation-building, Identity and War Commemoration Spaces in Malaysia and Singapore

Kevin Blackburn

Introduction

The recent work of cultural geographers on deathscapes, landscapes that are dedicated to the dead, such as cemeteries and war memorials, has shown that analysis of the ideological underpinnings of these landscapes can reveal much about the social construction of ideas of race and the nation. Lily Kong, a Singapore cultural geographer, in her literature review of the topic of deathscapes noted how these landscapes reflect political and the cultural interactions over the 'allocation of meanings to particular places' (Kong 1999, 2). War deathscapes in Singapore and Malaysia offer contrasting Southeast Asian contexts in which to analyse diverging racial and national identities over a shared historical past and how these identities have been constructed and expressed spatially.

Different approaches to nation-building and national identity in Malaysia and Singapore have created contrasting war deathscapes built to commemorate the most traumatic historical event that the two countries have shared – the Japanese Occupation that occurred after the defeat of the British by the Japanese in World War Two. The greatest slaughter of this period occurred in 1942 when tens of thousands of civilians, mainly Chinese, were massacred by the Japanese in what came to be called the *sook ching* (meaning to 'purge' or to 'cleanse') massacres, designed to eliminate anti-Japanese elements of the population through indiscriminate killing. The spatial expressions of commemorating this horrific event in the Malaysian and Singapore landscapes are heavily influenced by how history and national identity are constructed in Malaysia and Singapore by the state and its ideology. In Malay dominated Malaysia, there is no national monument to the civilian war dead, only isolated and mostly small memorials found in Chinese cemeteries and Chinese parts of towns. In Chinese dominated Singapore, the 64.7 metre tall monument to the civilian war dead, who were almost all victims of the 1942 *sook ching* massacres, is in the heart of Singapore's civic district. This contrast suggests that the war deathscapes of Malaysia and Singapore offer insights into the interaction of cultural and political power relations that shape national identity. [1]

Massacre, Victimhood, and Chinese Identity

In 1942 after the fall of Singapore, the Japanese military began a series of atrocities against the Chinese population of Singapore and Malaysia, known as the *sook ching* massacres. These were designed to eliminate anti-Japanese elements in the Chinese community which had aided China in the Sino-Japanese War (1937–1945). On 17 February 1942, the Japanese commander of the 25th Army, Lieutenant General Yamashita Tomoyuki, gave the order to his subordinates for *genju shobun,* or 'severe punishment', of the Chinese population. From their experience fighting against guerrillas in China, Yamashita's subordinates interpreted the order to mean *shukusei*, (in Chinese *su qing* or *sook ching*) (WO 235/1004, 265–6, 621). This meant eliminating suspected opponents by 'purging' or 'cleansing' through execution without trial. It was commonly employed in China as indiscriminate killing of whole villages near where guerrilla activity had been reported. Japanese officers knew that innocent villagers would be killed in any *shukusei* operation.

The Japanese troops who captured Singapore and the Malay Peninsula brought with them from China their brutal methods in dealing with the Chinese, and the language of describing this treatment of Chinese civilians. In China, the Imperial Japanese Army often engaged in the slaughter of civilians in newly occupied areas to 'cleanse' or eliminate 'anti-Japanese' elements. However, in many rural areas, which the Japanese military did not have the troops to spare to permanently occupy, they did not have the time to accurately identify who was 'anti-Japanese', even though they made a show of screening civilians to try and identify guerrilla fighters. Indiscriminate killing thus occurred although there was a demonstration of identifying those who may be 'anti-Japanese' or guerrilla fighters (Li 1975, 187–216; Snow 1941, 348–56; Johnson 1963, 55–6, 207).

Many of the Japanese troops and their commanders, including Yamashita, had served in the war against China, where the Imperial Japanese Army had committed many atrocities on Chinese civilians as a method of subduing the population. The three Japanese divisions occupying Singapore and the Malay Peninsula, the 5th Division, 18th Division, and Imperial Guards had been involved in *shukusei* operations in China (Dreyer 1995, 237; Frei 2004, 27–36). According to Iris Chang, 'the Japanese soldier was not simply hardened for battle in China; he was hardened for the task of murdering Chinese combatants and noncombatants alike'.

---

1   Until 1965, Singapore together with Malaysia formed the Federation of Malaysia established on 16 September 1963 from former British colonial territories. This arrangement superseded the Federation of Malaya established on 31 August 1957. Though both Malaysia and Singapore were inextricably entwined historically, politically, culturally and economically, Singapore was 'expelled' or 'separated' from the Federation on 9 August 1965 to become an independent, sovereign state.

The Japanese military used live Chinese civilians and unarmed prisoners of war in target practice and macabre killing games to desensitise and 'to numb its men' in order for them to more easily kill Chinese civilians (Chang 1997, 55; Ienaga 1978, 166–8). Hayashi Hirofumi, a Japanese historian specialising on the *sook ching* massacres and war crimes, has noted that in the Sino-Japanese War (1937–1945) *genju shobun* was imported into northern China from its previous practice in Japanese occupied Manchuria by the commander of the *kempeitai* (military police) in China from 1938 to 1939 who had been a military advisor in Manchuria. Hayashi concludes that it was then adopted by the chief of staff of the Japanese North China Area Army, Yamashita Tomoyuki (Hayashi 1999, 41).

The singling out of the Chinese community from the other ethnic communities of Singapore and the Malay peninsula for systematic massacres had a horrendous affect on the Chinese memories of the Japanese Occupation compared to the other communities. Chen Su Lan, a Chinese community leader and member of the British Military Administration's advisory council, in his writings estimated that perhaps up to 100,000 Chinese were killed in both Singapore and on the Malay Peninsula because of these operations that stemmed from Yamashita's order in 1942 (Chen 1969, 217–22). This high figure was based on his interviews with survivors of massacres, stories from relatives of those taken away by the Japanese, and examinations of bodies recovered from massacre sites. It is difficult to put a figure on how many died because no accurate records were kept. Even in Singapore there is no reliable figure. Japanese official estimates given to the war crimes trial of 1947 into the Singapore *sook ching* massacre are 5,000, which is too low from the number of bodies recovered after the war, to 50,000, given by Chen and the Chinese press at the time of the 1947 trial, which is too high from the number of bodies recovered. Chen argued that another 50,000 were killed on the Malay Peninsula (WO 235/1004, 54; Chen 1969, 217).

Yamashita's 17 February 1942 *genju shobun* command was applied, not just in Singapore, but throughout the whole Malay Peninsula, which was occupied by the 25th Army until 1943. Massacres of the Chinese population started during February in Singapore and worked their way up the Malay Peninsula over the following months. Japanese officers used the order *genju shobun* to indiscriminately kill members of the Chinese community in the large cities of Singapore and Penang and in the small villages of Malaysia (Chin 1946, 105). Members of the Penang *kempeitai* (military police) rounded up Chinese civilians in the first Penang *sook ching* of April 1942 to screen then kill them. In between Singapore's February and Penang's April *sook ching* massacres, Yamashita's command of *genju shobun* worked its way through the Japanese military along the west coast of the Malay Peninsula, which was where many Chinese towns and villages were located. On 28 February 1942, 2,000 Chinese were slaughtered in the town of Kota Tinggi, Johor. Chen mentioned that he recorded eyewitnesses saying that 'children were thrown into the air and fell on the bayonets held to receive them and carried up on bayonets in a procession through the town for fun. Pregnant women had their bellies split open for fun' (Chen 1946). On 4 March, over 300 Chinese civilians were massacred at the Chinese village of Gelang Patah, in the southwest district of Pontian in Johor. Nearby at Benut, a town of 1,000 people, on 6 March 'the men were packed in the market and the women in a Malay school, who were raped, and all were slaughtered' (see Figure 6.1). Johor became a horrific killing field of the

Figure 6.1   Major Chinese massacres in the Malay Peninsula, 1942

Source: Map by author, Kevin Blackburn.

Chinese during late February and early March 1942. There were major massacres in Johor Bahru, Senai, Kulai, Sedanak, Pulai, Rengam, Kluang, Yong Peng, Batu Pahat, Senggarang, Parit Bakau, and Muar. The Chinese postwar press estimated that in Johor alone 25,000 Chinese civilians were massacred (Chen 1969, 213–16).

By mid-March, Yamashita's *genju shobun* order was reaching the states of Malacca and Negeri Sembilan, on Johor's northwest border. On 16 March, the Malacca *kempeitai* executed on a remote beach outside of the town at Tanjong Kling, 142 Chinese civilians who they had with the help of informers arrested and detained in prison for being involved in aiding the China Relief Fund, the Overseas Chinese movement meant to fund China's resistance to Japan (WO 325/85). On 15 March, 76 Chinese males were killed at Kuala Pilah in Negeri Sembilan. On the following day in nearby Parit Tinggi, the whole village of over a hundred Chinese was massacred (WO 235/1071). On 18 March, another complete village, this time of 990 Chinese civilians, Joo Loong Loong (now known as Titi), was massacred by the *kempeitai* Major Yokokoji Kyomi and his troops when he implemented Yamashita's order of *genju shobun* (WO 235/1096). After the massacre, the village of Joo Loong Loong disappeared. The progression of Japanese massacres of Chinese civilians culminated in the Japanese *kempeitai* chief in Penang, Major Higashigawa Yoshinoru receiving on 15 March the *genju shobun* command as relayed to him from his counterpart Lieutenant Colonel Oishi Masayuki, who as the *kempeitai* chief in Singapore had just finished overseeing the butchering of thousands of Chinese male civilians in Singapore in its *sook ching*. Higashigawa proceeded to round up, with the help of Chinese informers, in the first week of April, several thousand Chinese who were then imprisoned, tortured and finally killed in the following months (WO 235/931, 383–93).

Sook ching operations continued in earnest throughout 1942, and were a reoccurring nightmare for Chinese villagers during the Japanese Occupation. Whenever there was anti-Japanese guerrilla activity, the *kempeitai* and Japanese soldiers would launch a *sook ching* operation and indiscriminately massacre the Chinese civilians, even whole villages, such as Sungei Lui, a Chinese village of 400, that was, like Joo Loong Loong, in August 1942 wiped off the map by a *sook ching* operation in 1942 (WO 235/1070).

The horrendous process of *sook ching* strengthened the Chinese identity of the Chinese living in Singapore and Malaysia. Wang Gungwu, the eminent historian of the Overseas Chinese, has made the observation that in Singapore and Malaysia, 'the Japanese made no distinction whether a Chinese was a patriot or not; all Chinese faced the same kind of terror and fear.' He noted that 'the effect of this on the collective memory consolidated the sense of Chinese nationalism and forced a Chinese cultural identity on everyone, no matter how long the Chinese had been in the country' (Wang 2000, 17). A strong sense of being Overseas Chinese arose as the Chinese community in Malaysia and Singapore was singled out from the Malay and Indian communities and suffered *sook ching* massacres at the hands of the Japanese, just like their counterparts in China, and due to their connection to China's resistance to Japan. Resistance against the Japanese in Singapore and Malaysia was seen by the Chinese as part of broader resistance to the Japanese in China (Blackburn and Chew 2005). Many Chinese in Malaysia and Singapore already saw themselves as having China as their first home and the country they were living as their second home (Hara 2003, 13). The feeling of victimhood because they were Chinese only strengthened orientation towards China during the war and in the immediate postwar years.

The deep feeling of belonging to China that influenced the identity of the Chinese community of Malaysia found expression in the deathscapes that were created by them when they commemorated the experience of *sook ching* under the Japanese in the immediate postwar years. From 1946 to 1948, the Chinese communities of Malaysia that had experienced the *sook ching* massacres exhumed the bodies and placed them in mass graves at the site or in nearby Chinese cemeteries if it was an isolated area. In May 1947, the Negeri Sembilan Chinese Relief Fund Committee, which was originally set up in the 1930s to send money to help China against Japanese invasion, organised a central committee of 29 people, with 15 subcommittees for remote outlying areas to exhume and rebury the estimated 5,000 Chinese massacred in *sook ching* operations (*Straits Times*, 28 May 1947).

Throughout 1947, the Chinese Chamber of Commerce for Malacca sponsored the creation of the Malacca War Victims Memorial Committee which exhumed and reburied hundreds of victims of the *sook ching* massacres in the state. The committee collected remains of victims from various districts including those who had died on the beach at Tanjong Kling and had been active in the Kuomintang movement and the China Relief Fund, which helped Chiang Kai-shek's Kuomintang government resist Japan. They were reburied under a large memorial erected in the Bukit China, the Chinese cemetery of the town of Malacca. On top of the memorial was sculptured a large white 12 ray sun against a blue background. This was the flag of the ruling Kuomintang Party of China. The image also appeared on China's national flag under the Kuomintang from 1928 to the Communist takeover on 1 October 1949. Chiang Kai-shek, the President of China and Generalissimo of the Chinese armed forces, was even asked to compose an epitaph for the monument, which he did, writing in Chinese, 'Models of Loyalty and Virtue' (*Malay Mail*, 17 October 1947). The memorial was unveiled on 5 April 1948 by Sir Edward Gent, the High Commissioner of the Federation of Malaya (as Malaysia was called prior to 1963). In attendance were Tan Cheng Lock, the President of the Malayan Chinese Association, representing the Chinese community of Malaya, and Ng Pah Seng, China's Consulate General to Singapore. The memorial was a tangible expression of Overseas Chinese identity, demonstrating loyalty to China.

The epitaphs on the memorials to the Chinese dead reflected the early postwar Chinese literature on the experience of the community in Malaysia and Singapore, such as the China Relief Fund publication, *Da Zhanzheng Yu Nan Qiao: Malaiyade Zhi Bu* [The Great War and the Overseas Chinese: Malayan Section] published in December 1946 to January 1947. Chinese description of massacres and anti-

Japanese resistance emphasised that those Chinese who were killed were martyrs who had died at the hands of brutal Japanese invaders for *zuguo*, the 'fatherland' or the 'motherland' – China (see Li Tie Min 1947). Their victimhood was entwined with patriotism and loyalty to China. Lim Pui Huen, a Malaysian Chinese historian, in her work on the impact of the war in Malaysia has remarked that 'Japanese brutality has completely coloured the Chinese memory of war, but at the same time the collective memory of suffering has contributed to the Chinese sense of community' (Lim 2000, 153).

The public war memorials descriptions of the victims of the *sook ching* regularly use the words *nan qiao* ['southern sojourners'] or *hua qiao* ['Chinese sojourners']. This reflects a similar trend in the 1940s Chinese literature on the war which uses the same language (see Li Tie Min 1947). The deliberate use of *qiao* ['sojourner'] to describe the Chinese in pre-independence Malaysia and Singapore indicated a strong attachment to China as home (Fitzgerald 1972, x). Cheah Boon Keng, a Malaysian Chinese historian, has concluded that 'on the whole, the Japanese occupation and the war experience strengthened Chinese nationalism and their sense of ethnic identity' (Cheah 1987, 54).

The *sook ching* memorial and commemorative space which most expresses Chinese nationalism and identity is in Johor Bahru. The town's Chinese community in August 1947 built Malaysia's largest *sook ching* memorial to 2000 Chinese civilians massacred around the town in 1942. Their remains were buried under a large monument located outside of the centre of the town and next to the cemeteries along Jalan Kebun Teh. The memorial is adjacent to the Foon Yew Chinese school; both being built on land owned by the Chinese community (see Figure 6.2).

The Johor Bahru *sook ching* memorial was constructed so that mourners paying their respects on commemorative occasions, such as *qing ming*, when the Chinese visit their ancestors' graves to pay their respects, enter the memorial park through a *pai lou*, a three arched gateway. The *pai lou's* three arches represent the three islands of immortality in Chinese culture. The gate is decorated with symbols and prayers to protect its entrance and ward off evil spirits from the mass grave of the 'martyrs' as the victims were called in Chinese. This type of gate is common in Chinese culture. From the Song Dynasty onwards, the Chinese used these gates to protect the entry into places of commemoration, such as tombs, graves of heroes, and lords. Even in Republican China, Sun Yat Sen's mausoleum has such a gate, as does that of Chiang Kai-shek in Taiwan. Many overseas Chinatowns have a *pai lou* at their entrance. Singapore's Chinese language university Nantah had a large *pai lou* at its entrance when it was founded. These gates thus also function so as to mark Chinese identity and entry into Chinese community space (Ong 2006, 116; Wan 2006). Interestingly, the large Ayer Hitam memorial to the *sook ching* victims in the northwest of Johor, which was built in 1947 outside of the town area, also has its own *pai lou* with prayers and protective imagery, such as dragons, to ward off evil spirits and to delineate the area as Chinese community space.

To further emphasise Chinese identity, some *pai lou*s at the *sook ching* memorials have modern Chinese political symbols. In August 1947, the *pai lou* at

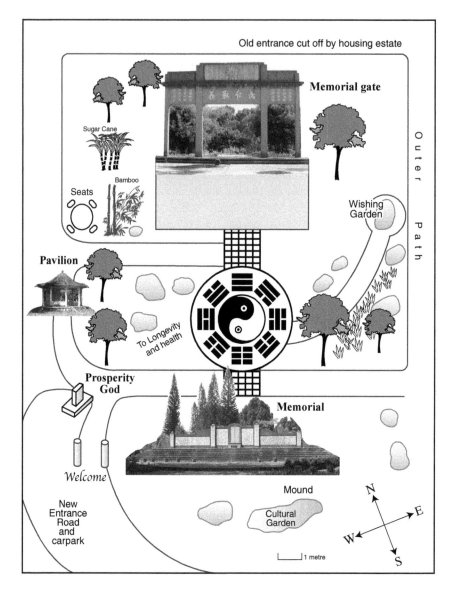

Figure 6.2   Memorial to Chinese War Dead in Johor Bahru

Source: Map by author, Kevin Blackburn

the Johor Bahru *sook ching* memorial was even emblazoned at its top with a large white 12 ray sun of the Kuomintang to demonstrate the Chinese community's identity as *nan qiao*, whose first home was China. In 1946 at Mentakab, a Chinese town in Pahang, the memorial to the Chinese war dead was built with a three arch way *pai lou* at its entrance. On the monument is the white 12 ray sun of the

Kuomintang (Franke and Cheng 1985, 493). This political imagery reflects the Chinese community's belief in the 1940s that they owed their political allegiance to China, and in particular the ruling party of China, the Kuomintang.

At the Johor Bahru *sook ching* memorial, the victimhood of the Overseas Chinese was couched in traditional Chinese metaphors and legend. The memorial has inscribed on it 51 names of the victims of the Chinese patriotic China Relief Fund who were killed by the Japanese. There are Chinese verses on the memorial composed by Li Szu-yuan which compare the Japanese invaders and the Chinese they massacred in Malaysia, Singapore and China to the massacres of Chinese civilians by the invading Manchus at Yang-chou and Chia-ting in China during 1645 when they established China's Qing Dynasty. Thus the memorials were contextualised within a Chinese perception of victimhood drawn from Chinese culture and history. This feeling of victimhood of the Overseas Chinese in Malaysia as a source of identity drew strongly upon the sense of victimhood that was entwined with nationalism in China itself. In twentieth century Chinese history, etched into the collective memory of the Chinese is the impression that in the fight against Japan during World War Two, the Chinese were victims enduring suffering and misery at the hands of the brutal Japanese (Dittmer and Kim 1993, 304). The memorials and commemorative spaces the Overseas Chinese community erected to the victims of the *sook ching* massacres reflected Chinese nationalism and were communal places – Chinese community space. The ideological underpinnings of the monuments were shaped by Chinese nationalism present in the identity of the Overseas Chinese.

War Deathscapes and National Identity in Malaysia

The Chinese were just one ethnic group in Malaysia and Singapore. Migration to Malaysia and Singapore meant that in the postwar years there were three main ethnic groups in both countries – Malays, the indigenous population; the Chinese and the Indians (see Tables 6.1 and 6.2). Nationalism among the other ethnic groups of Malaysia and Singapore, principally the Malays and Indians, was also strong after the war, but although they suffered during the Japanese Occupation they did not share the wartime experience of the Chinese as being victims of large scale massacres. The path to nationhood in the postwar years, and then afterwards nation-building, in Malaysia and Singapore presented problems with monuments that were linked to a strong Overseas Chinese identity. Diverging identities were reflected in the deathscapes. It is worth examining in detail how the nationalism of the other ethnic groups was at variance with the Chinese sense of nationalism that was based on victimhood and was memorialised in the 1940s. In the era of attaining nationhood and independence, then afterwards nation-building, these diverging identities presented problems with representing the wartime experiences.

For the Malay and Indian communities of Malaysia and Singapore, the Japanese Occupation had awakened strong nationalist feelings. Rather than the period being remembered as a time of unrelenting torture and massacre, members

Table 6.1   Percentage distribution of population by race in Malaysia

Year	Malay	Chinese	Indian	Others
1947	49.5	38.4	10.8	1.3
1957	49.8	37.2	11.3	1.8
1970	53.2	35.4	10.6	0.8

Table 6.2   Percentage distribution of population by race in Singapore

Year	Malay	Chinese	Indian	Others
1947	12.1	77.8	7.4	2.8
1967	14.5	74.4	8.1	3.0

Sources: Malaysian and Singapore Census, compiled by author, Kevin Blackburn.

of the Malay and Indian elites viewed it as time of national awakening on the road to independence. The Japanese had used a policy of divide and rule with the different ethnic groups. They had fanned fledging nationalism by fostering the desire for independence among the Malays and persuaded the Indian population that Japan would help Indians liberate India from the British.

The Japanese wanted the support of the Malays, so they encouraged Malay nationalist leaders, such as Ibrahim bin Yaacob, Mustapha Hussein, Ahmad Boestamam, Dr Burhanuddin Al-Helmy, and Ishak bin Haji Mohammad of the *Kesatuan Melayu Muda* [KMM or Union of Malay Youth] (Mustapha Hussain 2005). Members of the KMM welcomed the Japanese as liberators from British colonialism and assisted them through fifth column activities. Although the Japanese preached the propaganda of 'Asia for Asians', they had no intention of giving the Malays independence, and overtime the Malays grew to hate the Japanese as many suffered starvation and ill-treatment. Malays started to join resistance groups. Among them were Abdul bin Razak, the future Malaysian Prime Minister, Abdul Rahman bin Haji Talib, and Yeop Mahidin bin Mohammed Shariff. The KMM was banned by the Japanese when it demanded immediate independence. Still Malay nationalists made the most of the co-operation that Japan extended to them. The Japanese created the volunteer armies of mainly Malays with two KMM members at their head, Ibrahim bin Yaacob as head of the *Giyu Gun* and Onan Haji Siraj as head of the *Giyu Tai*. The Malay nationalists saw these armies as future nationalist armies that might be used to gain independence one day (Mustapha Hussain 2005, 257). The Japanese also sent Malays back to Japan for training and gave them many opportunities that the British had denied them.

Abu Talib Ahmad, a Malaysian historian, makes the point that members of the Malay elite who seized these opportunities for education and development in running the country given during the Japanese Occupation, as a postwar elite took the view that the Japanese Occupation was about a nationalist awakening; and they as an elite tended to make this the 'official view' of Malaysia (Abu Talib Ahmad 2003 and 2006). Abdul Aziz Ishak, a former KMM member, for example, served in the 1955 and post-independence 1957 cabinets under Malaysia's first Prime Minister Tunku Abdul Rahman. Abu Talib Ahmad quotes the Kuala Lumpur based Museum of National History's description of the period in Malay: 'Although the Japanese occupation brought suffering to the people, it also, to a certain extent, brought about a new awareness among them that led them to reconsider the capability of British colonial power and increased their struggle for independence' (Abu Talib Ahmad 2006, 40).

Malay representation of the wartime experience of the Japanese Occupation is thus as at variance with that of the Chinese. Senior politicians in the United Malays National Organisation (UMNO) the Malay ruling party of Malaysia, have expressed these sentiments. Abdullah Ahmad voiced them as editor in chief of the major English language daily in Malaysia, the *New Straits Times* (2001–2003), in an opinion piece marking sixty years since the beginning of the Japanese Occupation. Abdullah Ahmad noted that the Japanese Occupation proved that 'there was nothing pre-ordained about subjugation as the "white man's burden"'; and because the Japanese 'were Asian, like us … they induced the germ of an aspiration' that meant 'for my generation the ending of the Occupation was the beginning of nascent Malay nationalism' (*News Straits Times*, 18 December 2002).

The view of the wartime past as a time of emerging political consciousness and rising Malay nationalism that was facilitated by Japan has become the national view of Malaysia through the dominance of the Malay ruling party, UMNO. In August 1994, Prime Minister Mahathir Mohammad of Malaysia urged his Japanese counterpart Murayama Tomiichi to 'stop apologising for wartime crimes committed about 50 years ago' (*Straits Times*, 28 August 1994). Mahathir's view was completely at odds with that of the Chinese community of Malaysia, but very much in accord with the view that the Japanese Occupation was a time heightened political consciousness and developing Malay nationalism, which was not something that Japan should apologise for.

Malaysian national history as represented in school textbooks has reproduced this view. In the 1990 and 2004 revised secondary history curriculum for Malaysian schools, there is no mention of the *sook ching* massacres. However, there are prominent accounts of how the Japanese Occupation strengthened Malay nationalism through encouraging leaders such as those of the KMM, like Ibrahim Yaacob, with policies of 'Asia for Asians' (Sabihah Osman, Muzaffar Tate, and Ishak Ibrahim 1990, 11–12; Ramlah binti Adam, Abdul Hakim bin Samuri and Muslimin bin Fadzil 2004, 22–3; Siti Zurainia Abdul Majid, Muhammad Yusoff Hashim, Abdullah Zakaria Ghazali, Lee Kam Hing, Ahmad Fawzi Basri, and Zainal

Abdin Abdul Wahid 1992, 85–7; Ramlah binti Adam, Abdul Hakim bin Samuri, Shakila Parween binti Yacob and Muslimin bin Fadzil 2004, 56–9). Malaysian history taught in schools presents its national history as starting with the fifteenth century Malacca Sultanate of Malay culture, which is viewed as a golden age of Malay achievement. This was interrupted by European colonisation. The Japanese Occupation is seen as a period that awakens nationalism and starts the Malays, united under UMNO, towards gaining independence from the colonial powers and once again being able to emulate the golden age of the Malacca Sultanate. Thus it is not viewed as a time of the slaughter of innocent victims. The Malaysian historian Abu Talib Ahmad laments that ill-treatment of the Malays by the Japanese and thousands of poor Malay labourers who were sent to the deaths by being forced by the Japanese to work on the Burma-Thailand Railway are left out of this glorious nationalist narrative provided by the Malay ruling elite (Abu Talib Ahmad 2000, 81 and 2006).

Forgotten in Malaysian history, as told in the national education system, are the massacres of tens of thousands of Chinese civilians in 1942. However, these massacres are marked by dozens of well-looked after memorials geographically scattered all over Malaysia and tended by members of the Chinese clans in order to keep alive the memory of the victims. Many of these war deathscapes are in Chinese cemeteries and not civic or public areas. Commemorative space in Malaysia reflects the plural society of Malaysia in which the different Malay, Chinese, and Indian communities essentially live separate lives although they live amicably and peacefully alongside each other.

In its representation of history, the Malaysian state has not seriously upheld its sometimes stated goal of a *Bangsa Malaysia* (a united Malaysian race or nation). The Malay dominated state has been uncomfortable with anything outside having history and culture dominated by the idea of *Bangsa Melayu* (Malay race), which does not embrace the history of the other ethnic communities of Malaysia in national narratives (Kua 1990; Ariffin Omar 1993). Malaysian history textbooks actually assert the concept of Malay supremacy, *Ketuanan Melayu*, in Malaysian public life (*The Star*, 11 December 2004; Ramlah binti Adam, Abdul Hakim bin Samuri, Shakila Parween binti Yacob and Muslimin bin Fadzil 2004). The ethnic groups that do not control the state are left to themselves, without the sponsorship of the state, to find their own commemorative space. Thus remembering the war dead in Malaysia has become a Chinese affair as members of Chinese clans and chambers of commerce continue to attend commemoration ceremonies and maintain the monuments to the Chinese *sook ching* victims. The deathscapes are a spacial reflection of a Chinese identity.

In Malaysia, the wartime narrative of the massacres of the Chinese and their deathscapes could not be integrated into the national narrative of the wartime past because Malay dominance in public life clearly found it unacceptable to incorporate a remembering of the past that represented such a strong Chinese identity. It is worthwhile comparing the fate of Malaysia's war deathscapes associated with the *sook ching* to the creation of deathscapes in Singapore, which shared this wartime experience. Singapore, because of its dominant Chinese population, is unlikely to have pushed deathscapes of the *sook ching* massacres and commemoration of them to the periphery, as in Malaysia.

War Commemorative Space and Nation-building in Singapore

In Singapore, the state sponsored nation-building project of encouraging a 'Singaporean Singapore' identity has created a national war deathscape at the Civilian War Memorial, with a massive monument in the heart of the civic district of the city-state. Remembering the victims of the victims of the *sook ching* massacres has been transformed from just representing the Chinese community paying tribute to Chinese civilian war dead to a commemorative space that all ethnic groups pay homage to on 15 February each year – the anniversary of the fall of Singapore. Now it is an occasion in which all Singaporeans regardless of racial identity remember their collective suffering under the Japanese Occupation and are encouraged to draw the lesson that Singapore must remain united and accept military conscription in order to defend itself against potentially hostile neighbours who are irked by Singapore's multiracial identity based on meritocracy rather a society similar to their own in which one race dominates the state.

Malaysia went through a similar wartime experience as Singapore with tens of thousands of Chinese massacred, yet it does not have national commemorative spaces to mark the event. In Singapore, commemorative space dedicated to the *sook ching* war victims is designed to unite the nation. In Malaysia, similar commemorative spaces reflect the plural nature of the Malaysian identity, each minority community is left to commemorate its own experiences without the state creating a national war deathscape as the Singapore state has at the Civilian War Memorial.

The reasons why Singapore has a national commemorative deathscape to the victims of the *sook ching* massacres and Malaysia has communal monuments lies partly with changing identities. When the monuments in Malaysia were constructed during the 1940s the Chinese were much more closely connected to China. The independence struggle in Malaysia and Singapore had not begun in earnest. In the 1950s and 1960s, the closeness of the Chinese community to China would begin to diminish as they chose to become citizens of the new emerging countries of Singapore and Malaysia (Hara 2003) and cut themselves off from the Kuomintang and Communist China. British rulers, such Sir Franklin Gimson, the Governor of Singapore (1946–1952), encouraged the Chinese take up local citizenship and see themselves as belonging first and foremost to the country in which they lived in preparation for self-government and independence (CO 537/3758). The Chinese communities still continued with their ceremonies at the *sook ching* war deathscapes in Malaysia and renovated these monuments of the 1940s. In 1972, a renovation committee was formed for the upgrade of the Malacca monument. On 5 April 1993, *qing ming* day, when Chinese traditionally clean their ancestors' graves, the new upgraded monument with considerable additions was unveiled.

However, overtime, the connection with China that these monuments represented in the 1940s had faded considerably. They were much more an occasion for marking Malaysian Chinese identity, not an identity as *hua qiao*, or Overseas Chinese.

In Malaysia, the *sook ching* monuments were erected in the 1940s when the Chinese strongly saw themselves as *hua qiao*. In Singapore, the construction of a similar monument was delayed twenty years, and built in the context of creating a Singapore national identity after self-government in 1959 and full independence in 1965. In the aftermath of the Japanese Occupation, the Singapore Chinese Chamber of Commerce and other Chinese organisations moved to set up a committee to build a monument for all the Chinese civilians who had lost their lives in the *sook ching* massacre that took place in the whole of Singapore. In 1946, the Chinese Chamber of Commerce formed a committee to find and exhume the bodies of the *sook ching* victims, the 'Singapore Chinese Massacred Appeal Committee'. However, there was disagreement in the large Chinese community of Singapore over how to go about the exhumation of the victims and then rebury them (Shu Yun-Tsiao and Chua Ser-Koon 1984, 80, 84). The executive secretary of the 'Singapore Chinese Massacred Appeal Committee', Colonel Chuang Hui-Tsuan proved too divisive a figure to unite the myriad of Singapore Chinese clan and business organisations behind him to first exhume the remains of the *sook ching* victims, then to find a suitable place to bury them (Shu Yun-Tsiao and Chua Ser-Koon 1984, 35, 80–85). The process of identifying the locations of the remains of the victims and choosing a site for reburial dragged on without resolution. Finally, in 1962, when quarrying for sand disturbed the large *sook ching* mass war grave at Siglap, Colonel Chuang's committee's right to exhume the *sook ching* victims was withdrawn by the Singapore government and handed directly over to the Chinese Chamber of Commerce (*Nanyang Siang Pau*, 19 and 25 July 1962).

In charging the Chinese Chamber of Commerce and its president, Ko Teck Kin, with the responsibility of exhuming and reburying the *sook ching* victims, the Singapore government imposed conditions that any monument erected should reflect the nation-building project that itself had embarked upon, creating a 'Singaporean Singapore' rather than a Singapore polarised by strong ethnic identities. Ko and the Chinese Chamber of Commerce arranged for the involvement of other ethnic groups in the erection of a memorial, and it became a monument not just to the Chinese civilians killed during the war, but all civilians killed. Lee Kuan Yew, the Prime Minister of Singapore and leader of its ruling People's Action Party (PAP) government, later wrote that Ko 'knew that the PAP government would be unhappy as long as it was a purely Chinese issue', and therefore turned it into a national issue with a multiracial committee composed of chambers of commerce members of the Malay, Indian, Eurasian, and Ceylonese communities (Lee 1998, 496). On 15 June 1963, when turning the sod for the monument in the heart of Singapore's civic district, Lee Kuan Yew emphasised that the monument represented the sufferings of all the citizens of Singapore. He told the gathering, the Japanese Occupation was a 'cataclysm experience' that 'none of us here in Singapore will ever forget'. He reminded the crowd that 'part of the agony was the sudden disappearance of tens of thousands of our young men, mostly civilians, and some volunteers. Most were Chinese, but there were also Indians, Malays, Eurasians, Ceylonese and others. Even two Sikh families were massacred' (Republic of Singapore 1962–1963).

The involvement of all of Singapore's ethnic groups produced a deathscape with a monument and ceremonial practices that were national rather than ethnicised. The monument itself from 1963 was called the Civilian War Memorial (Figure 6.3). The government allocated land appropriate for what was considered a national monument. It was in the heart of the Singapore's civic district, near the

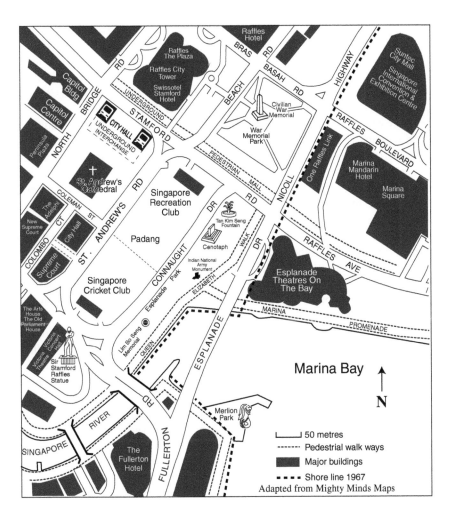

Figure 6.3   Singapore's Civic Centre and Civilian War Memorial

Source: Map adapted by author, Kevin Blackburn.

government buildings, such as Parliament, City Hall, and the Supreme Court. It was close to other national monuments, such as the Cenotaph to Allied soldiers killed in the war, the memorial to war hero Lim Bo Seng and the Raffles statue. The plot of land allocated to the Civilian War Memorial was in front of the elite school Raffles Institution and in the 1960s its 64.7 metre tall pillars dominated the skyline overlooking the sea. Underneath the moment were stored 660 urns of the bones of the thousands of victims exhumed from the mass graves.

Unlike the traditional Chinese *sook ching* monuments of Malaysia, the Civilian War Memorial represents a national identity that moulds the diverging ethnic identities into one. It consists of four pillars which represented the suffering of the four main communities of Singapore, the Chinese, Malay, Indians, and Eurasians. When the Civilian War Memorial was unveiled by Prime Minister Lee Kuan Yew on 15 February 1967, it was proclaimed that the memorial 'consists of four tapering columns' meant 'to symbolise the four streams of culture merging into a single entity' (*Straits Times*, 16 February 1967). Inscriptions on the platform supporting these pillars appear in Chinese, Malay, and English. The original proposed Chinese inscription describing the memorial to the victims of the *sook ching* massacres in Singapore was scrapped as it did not fit into the theme of all ethnic groups enduring suffering (Pan Shou 1984, 57–60). The Chinese emblems and symbols present at the memorial to the *sook ching* victims of Johor Bahru contrast strongly with the stark naked nature of the Civilian War Memorial, which has no visual imagery that links the monument to Chinese culture. *Feng shui* (literally meaning 'wind and water') of the Johor Bahru monument is very much in the Chinese classic style. Most Chinese war memorials in Malaysia are built in the desirable *feng shui* style of being on the side of hills overlooking water. The Civilian War Memorial was not built on a hill but in the civic district of Singapore as a national monument. In 1967, the Civilian War Memorial did look out over the sea, but since then it has been hemmed in by many tall buildings that the Singapore government has allowed to be built around it. It has bad *feng shui*, but its position still befits that of a national monument in the centre of Singapore's civic district.

The national nature of the Civilian War Memorial of Singapore has been present in the rituals and ceremonies that have surrounded it since its inception. When unveiling the Civilian War Memorial on the twenty-fifth anniversary of the fall of Singapore, 15 February 1967, Prime Minister Lee affirmed that the Singapore nation had endured suffering together and the memory of this united it:

> This piece of concrete commemorates an experience which in spite of its horrors, served as a catalyst in building a nation out of the young and unestablished community of diverse immigrants. We suffered together. It told us that we share a common destiny. And it is through sharing such common experiences that the feeling of living and being one community is established (Republic of Singapore 1967).

At the ceremony to unveil the Civilian War Memorial, state sanctioned rituals marked the monument as part of national public space rather than an ethnicised entity. Prayers were said by leaders of the Inter-Religious Council representing the Muslim, Buddhist, Christian, Hindu, Jewish, Sikh, and Zoroastrian faiths (Blackburn 2000, 79–90). However, the crowd attending was mainly comprised of large numbers of elderly Chinese women. According to Taoist rituals, they made offerings to their dead relatives to ensure that their relatives would not want in the after-life or become 'hungry ghosts' condemned to dwell in hell because they had violent deaths and their remains had never been recovered so their children could tend to them and make offerings.

The ceremonies held at Singapore's Civilian War Memorial since 1967 although organised by the Chinese Chamber of Commerce have included individuals and organisations from both national bodies and as representatives of each ethnic group. In the early remembrance ceremonies, members of the Malay and Indian Chambers of Commerce turned out prominently to join their Chinese counterparts at the Civilian War Memorial on every 15 February (*Straits Times*, 16 February 1968 and *Eastern Sun*, 16 February 1969). Together they marked the memory of how the nation suffered during a key watershed in Singapore's national history, as told in its school history. There was renewed focus on the ceremony at the Civilian War Memorial on 15 February when the date was designated Total Defence Day in 1998. Singapore school children began to attend in greater numbers than before. They were told to remember the atrocities of World War Two in order to understand more fully the National Education message: 'We must ourselves defend Singapore'. According the Singapore government, 'the day in 1942 when Singapore fell to the Japanese' will 'serve as a reminder that everyone has a part to play in the country's defence' (*Straits Times*, 18 May 1997).

The fall of Singapore and the subsequent harsh period of the Japanese Occupation has certainly since the early 1990s been drawn upon in Singapore in Ministry of Education textbooks as 'a lesson of history' for defending Singapore (Wong 2001, 218–38). This is because modern Singapore is a country in which all males when they reach the age of eighteen must commence over two years of National Service, and then continue to come back regularly for more military training. To deter any potential future enemy, Singapore claims that it can bring 300,000 well-trained men under arms in 24 hours. The social studies textbook for secondary schools used since 1994 explicitly makes a connection between the failure of the British in defending the people of Singapore in 1942 from the Japanese and the need for National Service and military training in contemporary Singapore. The textbook says that 'from the British defeat we learn' that 'a country must always be well-prepared for any attacks from enemies' and that 'it must not depend on others to protect its people'. The textbook goes on to draw the lesson from life during the Japanese Occupation as being that 'the people must be trained to defend their own country'. Thus, 'in 1967, the government started National Service' in order 'to enable all young men to be trained to defend Singapore in case of war' (Curriculum Development Institute of Singapore 1994, 97).

The remembrance ceremony conducted at Singapore's Civilian War Memorial every 15 February reflects the lessons of history found in school textbooks. In recent years, on the morning of every 15 February, hundreds of children from different schools with their teachers gather around the memorial with various dignitaries. A senior cabinet minister from the Singapore government regularly attends the function as the guest of honour. First, they listen to the 'All Clear' signal sounded by the Singapore Civil Defence Force indicating that Singaporeans can come out of their shelters in the event of a bombing attack. The following speeches from community leaders attending remind the audience of the sacrifices of the nation and reaffirm the National Education message of 'We must ourselves defend Singapore'. Members of the Singapore military take up positions of guards of honour around the memorial. Prayers are then offered to the war dead from the Inter-Religious Council, followed by a minute of silence. Then wreaths are laid at the memorial. The first organisation to lay a wreath is the Chinese Chamber of Commerce, next is the Singapore Armed Forces Veterans' League, followed the by Inter-Religious Council. Members of the National Cadet Corps from Singapore schools then lay a wreath. Finally, children from different schools lay wreaths. The audience concludes the ceremony with three bows to the memorial, followed by the playing of the Last Post and a final minute of silence. While the event is organised by the Chinese Chamber of Commerce, it is clear that it is a national event not an ethnicised occasion. There are traditional Taoist offerings of fruit, paper money, burnt joss sticks; but they are on the periphery of events. In contrast, in Malaysia, these Taoist rituals are central to the Chinese community ceremonies at the memorials in Malaysia to the *sook ching* victims, which are very much ethnicised occasions taking place in Chinese community space – Chinese cemeteries (Li Ye Lin 1996, 365–410).

Comparing and contrasting deathscapes that are dedicated to marking the massacres of the Chinese in 1942 in Malaysia and Singapore offers insights into how public commemorative spaces are shaped by diverging identities. The experiences of the two Southeast Asian countries confirm Kong's assessment that such deathscapes have strong ideological underpinnings reflecting how ideas of race and national identity are constructed. In Malaysia, these commemorative spaces are clearly seen as Chinese community space and not national space. In Singapore, the reverse is true. Why this is so is because identity is constructed differently in each country. In the plural society of Malaysia, with Malay dominance in social and cultural life, and with each group essentially living separately from the others, the minor ethnic groups are left to themselves to mark their own history. In Singapore, there is a strong focus on making the experiences of the different ethnic groups part of the Singapore national identity through including them in the national historical narrative. The suffering of the dominant ethnic group in Singapore, the Chinese, during the Japanese Occupation could not be seen purely in ethnic terms but also had to include the suffering of other ethnic groups. Thus, despite Singapore and Malaysia having a shared past, their commemorative spaces to this same history mirror their diverging identities.

References

Abu T. Ahmad (2000), 'The Malay Community and Memory of the Japanese Occupation', in P. Lim Pui Huen and Diana Wong (eds), *War and Memory in Malaysia and Singapore* (Singapore: Institute of Southeast Asian Studies).

Abu T. Ahmad (2003), *The Malay Muslims, Islam and the Rising Sun: 1941–1945, MBRAS Monograph No. 34* (Kuala Lumpur: MBRAS).

Abu T. Ahmad (2006), 'Museums and the Japanese Occupation of Malaya', in Richard Mason and Abu Talib Ahmad, (eds), *Reflections on Southeast Asian History Since 1945* (Penang: Penerbit Universiti Sains Malaysia).

Ariffin Omar (1993), *Bangsa Melayu: Malay Concepts of Democracy and Community: 1945–1950* (Kuala Lumpur: Oxford University Press).

Blackburn, Kevin (2000), 'The Collective Memory of the *Sook Ching* Massacre and the Creation of the Civilian War Memorial of Singapore', *Journal of the Malaysian Branch of the Royal Asiatic Society* 73:2, 71–90.

Blackburn, Kevin and Chew, Daniel Ju Ern (2005), 'Dalforce at the Fall of Singapore in 1942: An Overseas Chinese Heroic Legend', *Journal of the Chinese Overseas* 1: 2, 233–59.

British Military Administration, Malaya. Singapore Advisory Council, Report of Proceedings, Chen Su Lan, volume 1, 23 January 1946, p. 129.

Chang, Iris (1997), *The Rape of Nanking* (New York, Penguin).

Cheah Boon Keng (1987), *Red Star Over Malaya: Resistance and Social Conflict During and After the Japanese Occupation, 1941–1946*, 2nd edition (Singapore University Press).

Chen Su Lan (1969), *Remember Pompong and Oxley Rise* (Singapore: Chen Su Lan Trust).

Chin, Kee Onn (1946), *Malaya Upside Down* (Singapore: Jitts).

CO 537/3758 Political Developments: Chinese Affairs (The National Archives, London).

Curriculum Development Institute of Singapore (1994), *Social Studies: Secondary 1* (Singapore: Longman).

Dittmer, Lowell and Kim, Samuel S. (eds) (1993), *China's Quest for National Identity* (Ithaca: Cornell University Press).

Dreyer, Edward L. (1995), *China at War, 1901–1949* (Harlow, Essex: Longman).

Eastern Sun (Singapore).

Fitzgerald, Stephen (1972), *China and the Overseas Chinese: A Study of Peking's Changing Policy, 1949–1970* (Cambridge University Press).

Franke, Wolfgang and Chen, Tieh Fan (1982–1985), *Chinese Epigraphic Materials in Malaysia*, 3 volumes (Kuala Lumpur: University of Malaya Press).

Frei, Henry (2004) *Guns of February: Ordinary Japanese Soldiers' Views of the Malayan Campaign and the Fall of Singapore, 1941–1942* (Singapore University Press).

Hara Fujio (2003), *Malayan Chinese and China: Conversion in Identity Consciousness 1945–1957* (Singapore University Press).

Hayashi Hirofumi (1999), 'Japanese Treatment of Chinese Prisoners, 1931–1945', *Nature-People-Society: Science and Humanities* 26: 39–52.

Ienaga Saburo (1978), *The Pacific War: World War II and the Japanese, 1931–1945* (New York: Pantheon).

Johnson, Chalmers A. (1963), *Peasant Nationalism and Communist Power: The Emergence of Revolutionary China 1937–1945* (London: Stanford University Press).

Kong, Lily (1999), 'Cemeteries and Columbaria, Memorials and Mausoleums: Narrative and Interpretation in the Study of Deathscapes in Geography', *Australian Geographical Studies* 37:1, 1–10.

Kua Kia Soong (ed.) (1990), *Malaysian Cultural Policy and Democracy* 2nd edition (Kuala Lumpur: Selangor Chinese Assembly Hall).

Lee Kuan Yew (1998), *The Singapore Story* (Singapore: Times).

Li, Lincoln (1975), *The Japanese Army in North China, 1937–1941: Problems of Political and Economic Control* (Tokyo: Oxford University Press).

Li Tie Min (ed.) (1947), *Da Zhan Yu Nan Qiao: Malaiya Zhi Bu* [The Great War and and the Overseas Chinese: Malayan Section] (Singapore: Singapore Overseas Chinese Publishing Company).

Li Ye Lin (ed.) (1996), *Tai Ping Yang Zhan Zheng Shi Liao Hui Ban: Selected Historical Materials of the Pacific War* (Kuala Lumpur: Selangor Chinese Assembly Hall).

Lim, P. Pui Huen (2000), 'War and Ambivalence: Monuments: Monuments and Memorials in Johor', in P. Lim Pui Huen and Diana Wong, (eds), *War and Memory in Malaysia and Singapore* (Singapore: Institute of Southeast Asian Studies).

Malay Mail (Kuala Lumpur).

Mustapha Hussain (2005), *Malay Nationalism Before UNMO*, trans. Insun Sony Mustapha, Jomo. K.S. (ed.), (Kuala Lumpur: Utusan Publications and Distributors).

Nanyang Siang Pau (Singapore).

New Straits Times (Kuala Lumpur).

Ong Hean-Tatt (2006), *Scientific Statistical Evidence for Feng Shui* (Subang Jaya, Malaysia: Gui Management Centre).

Pan Shou (1984), 'Memorial to the Civilian Victims of the Japanese Occupation', *Journal of the South Seas* 39: 57–60.

Ramlah binti Adam, Abdul Hakim bin Samuri and Muslimin bin Fadzil (2004), *Sejarah Tingkatan 3: Kurikulum Bersepadu Sekolah Menengah* [History for Form 3: Secondary School] (Kuala Lumpur: Dewan Bahasa dan Pustaka).

Ramlah binti Adam, Abdul Hakim bin Samuri, Shakila Parween binti Yacob and Muslimin bin Fadzil (2004), *Sejarah Tingkatan 5: Kurikulum Bersepadu Sekolah Menengah* [History for Form 5: Secondary School] (Kuala Lumpur: Dewan Bahasa dan Pustaka).

Republic of Singapore: Prime Minister's Speeches, Press Conferences, Interviews, Statements, etc., 1962–1963, Prime Minister's Speech at 'Breaking of the Sod' for the Memorial to Civilian Victims During the Japanese Occupation, On 15 June, 1963 (National University of Singapore Library).

Republic of Singapore: Prime Minister's Speeches, Press Conferences, Interviews, Statements, etc., 1967, Prime Minister's Speech at the Unveiling Ceremony of the Memorial to Civilian Victims of the Japanese Occupation On 15 February, 1967 (National University of Singapore Library).

Sabihah Osman, Muzaffar Tate, and Ishak Ibrahim (1990), *Sejarah Tingkatan 3: Kurikulum Bersepadu Sekolah Menengah* [History for Form 3: Secondary School] (Kuala Lumpur: Dewan Bahasa dan Pustaka).

Shu Yun-Tsiao and Chua Ser-Koon (eds) (1984), *Malayan Chinese Resistance to Japan 1937–1945: Selected Source Materials Based on Colonel Chuang Hui-Tsuan's Collection*, (Singapore: Cultural and Historical Publishing House).

Siti Zurainia Abdul Majid, Muhammad Yusoff Hashim, Abdullah Zakaria Ghazali, Lee Kam Hing, Ahmad Fawzi Basri, and Zainal Abdin Abdul Wahid (1992), *Sejarah Tingkatan 5: Kurikulum Bersepadu Sekolah Menengah* [History for Form 5: Secondary School] (Kuala Lumpur: Dewan Bahasa dan Pustaka).

Snow, Edgar, *The Battle for Asia* (New York: World Publishing Company)

Straits Times (Singapore).

The Star (Kuala Lumpur).

Wan Meng Hao (2006), Executive Secretary of The Preservation of Monuments Board, personal communication.

Wang Gungwu (2000), 'Memories of War: World War II in Asia' in P. Lim Pui Huen and Diana Wong (eds), *War and Memory in Malaysia and Singapore* (Singapore: Institute of Southeast Asian Studies).

Wong, Diana (2001), 'Memory Suppression and Memory Production: The Japanese Occupation of Singapore', in T. Fujitani, Geoffrey M. White and Lisa Yoneyama (eds), *Perilous Memories: The Asia-Pacific War(s)* (Durham, North Carolina: Duke University Press).

WO 235/931 – Defendant: Higashigawa Yoshinoru: Place of Trial Penang (The National Archives, London).

WO 235/1004, Defendants: Nishimura Takoma, Kawamura Saburo, Oishi Masayuki,Yokota Yoshitaka, Jyo Tomotatsu, Onishi Satoru, and Hisahatsu Haruji, Place of Trial: Singapore (The National Archives, London).

WO 235/1070, Defendant: Hashimoto Tadashi: Place of Trial Kuala Lumpur (The National Archives, London).

WO 235/1071, Defendants: Watanabe Tsunahiko, Iwata Mitsugi, and Goba Itsugi: Place of Trial Kuala Lumpur (The National Archives, London).

WO 235/1096, Defendant: Yokokoji Kyomi: Place of Trial Kuala Lumpur (The National Archives, London).

WO 325/85 Malacca, Malaya: killing and ill treatment of Asians (The National Archives, London).

Chapter 7

Being Javanese in a Changing Javanese City

Ambar Widiastuti

Globalisation

The word 'globalisation' might be one of the most used political terms today beside 'terrorism'. It is like a ghost: you are able to feel its presence but yet it is difficult to define. Anthony Giddens defines globalisation as 'the intensification of worldwide social relations which link distant localities in such a way that local happenings are shaped by events occurring many miles away and vice versa' (Giddens 1990, 64). He, then, continues that 'this is a dialectical process because such local happenings may move in an observed direction from the very distanciated relations that shaped them'. Malcolm Waters defines globalisation as 'a social process in which the constraints of geography on social and cultural arrangements recede and in which people become increasingly aware that they are receding' (Waters 1995, 3). From the two definitions above, there are at least two variables of globalisation: 'interconnectedness' and 'awareness' with thanks to the advancement of communication and transportation technology which allows this to happen.

It is misleading to say that globalisation equals 'Westernisation' or 'Americanisation'. Nevertheless, the two terms may be seen as aspects of globalisation as one recognised the globalising phenomenon of industrialisation as beginning in the West. Industrialisation brought with it values like individualisation, universalism, secularity, rationalisation (Waters 1995, 13; Friedman 1994, 84), effectiveness and efficiency (Anshory and Thoha 2005, 241) under the umbrella of 'modernisation'. What then happens when these traits encounter Javanese culture in a Javanese city named Yogyakarta, on the island of Java: the fifth largest and most populous island constituting the Republic of Indonesia.

On Being a Javanese

What is it like being a Javanese in Yogyakarta today? Yogyakarta has always been an important part in Indonesia's political life. Shortly after Indonesia declared its independence in 1945, the Yogyakarta Sultanate declared their territory as part of this new Republic. During the Indonesian National Revolution (1945–1950) against the colonial Dutch rulers, the country's capital was moved to Yogyakarta due to the political instability in Jakarta. In return for this tremendous contribution and unfailing support, Yogyakarta was granted a 'Special Region' status, recognising the authority of the Sultan in his own region's domestic affairs. During the Reformation movement in 1998, Yogyakarta again and especially the Sultanate played an important role. The Sultan himself led a peaceful rally with estimated attendance of tens of thousands of people to protest the New Order regime. The relatively calm social and political situation in Yogyakarta, unlike other major cities in Indonesia during that phase, was believed to be influenced by the strong leadership of the Sultan: a phenomenon noted by other Indonesian cities. The Javanese, constituting the largest ethnic group in Indonesia, have perceptible political and social influence on the country from political and government directives to the school curriculum.[1] Events affected by and originated from the Javanese population would have an impact on the rest of the Republic.

Significantly, what is the importance of being a Javanese in Yogyakarta today? Known as the 'Students' City' of Indonesia, Yogyakarta has been 'invaded' by people and ideas from all around the archipelago and the rest of the world: thanks to intensive exposure from the mass media including the internet. Modern malls and clubs are being built, bringing and offering hedonistic metropolitan lifestyle leaving traditional cultural activities for the tourists. Western-like television shows are broadcast daily while local cultural programmes are not given enough airtime except on local television stations. These programmes are seemingly popular with the elders and those who live in the villages with the perception that these channels are not attracting other demographic groups of the population.

The sense is that Javanese cultural identity amongst Yogyanese, especially among the younger generation is experiencing perceptible erosion. As used here, 'identity' refers to the subjective feelings and valuations of any people which possesses common experiences and one or more shared cultural characteristics. Usually, this includes customs, language or religion. As noted by Anthony Smith (1990, 179), these feelings and values refer to three components of their shared experiences.

1. a sense of continuity between the experiences of succeeding generations of the unit of population;
2. shared memories of specific events and personages which have been turning-points of a collective history; and
3. a sense of common destiny on the part of the collectivity sharing those experiences.

Essentially, a collective cultural identity are those feelings and values with respect to a sense of continuity, shared memories and a sense of common destiny of a given unit of population which has had common experiences and cultural attributes (Smith 1990, 179). Thus, nurturing an identity is an enduring process.

---

1 Niels Mulder (2005) provides extensive elaboration on how Javanese principles are incorporated into the country's philosophy, forming what he calls 'the ideology of Javanese-Indonesian leadership'.

Yogyakarta

Yogyakarta is a unique province of Indonesia. It is the only monarchical province in the Republic and is formally governed by a pre-colonial Sultanate. The Yogyakarta Sultanate, formally the Sultanate of Ngayogyakarta Hadiningrat, was formed in 1755 when the existing Sultanate of Mataram was divided by the Dutch East India Company (*Vereenigde Oost-Indische Compagnie*) in two under the Treaty of Giyanti. This treaty stated that the Sultanate of Mataram was to be divided into the Sultanate of Ngayogyakarta Hadiningrat with Yogyakarta as the capital and Mangkubumi, who then became Sultan Hamengkubuwono I, as its Sultan. The other half formed the Sultanate of Surakarta Hadiningrat with Surakarta as the capital and Pakubuwono III as its Sultan. The ruler Sri Sultan Hamengkubuwono IX held a degree from Leiden University, Netherlands, and held the position of Vice-President of Indonesia (1973–1978), as well as Minister of Finance (1966–1973) and Minister of Defence (1949–1950, 1952–1953).[2] The current ruler of Yogyakarta is his son, Sri Sultan Hamengkubuwono X (born Bendara Raden Mas Herjuno Darpito), who holds a law degree from Gadjah Mada University, Yogyakarta.

In the past, the Sultan played a great role as the centre and guardian of Javanese culture. The Javanese believed, and some of them still do, that all good things came from the Sultan (Soemardjan 1982, 26). As Niels Mulder (2006) observed, the cultural setting of the Sultanate has eroded rapidly. The *Kraton* [palace] is no longer the dominant point of orientation for a social and political life. Since the Sultanate has always been considered a personification of Javanese culture, the implications of its perceived declining authority and the continual nurturing of Javanese culture are significant. Of course, these political and cultural concerns have echoes in other Southeast Asian cities and are not unique to Yogyakarta but the impact of globalisation on Java's historical, political and cultural legacies have ramifications for one of the largest sovereign countries in the world.

Predictably, one significant concern emerging from this perception or sense of cultural neglect can be seen in the use of Javanese language. Crucially, language is one of the aspects of ethnicity that helps to define the boundaries of a group (Diane and Mauzy 2002, 100). The Javanese language serves this function perfectly. In Javanese culture, there are two kinds of forms used: *kromo* [High Javanese] and *ngoko* [Low Javanese]. *Kromo* is the polite form of Javanese language, commonly used in speaking to social superiors, among the upper classes or in formal occasions (Mulder 2005, 38). This language imposes 'order', social control and economic mobility. It can regulate a person's station in life vis-à-vis others: conferring honour, governing expectations and codifying obligations (Mulder 2005, 45). *Ngoko* is common Javanese, used in speaking to social inferiors and among ordinary people and intimates (Mulder 2005, 38). In everyday life, younger Javanese tend to use *ngoko*, which is considered egalitarian, or the national Indonesian language of

---

2   <http://www.en.wikipedia.org/wiki/Yogyakarta> accessed 27 November 2007.

Bahasa Indonesia to address the elders. However in formal occasions, *Bahasa Indonesia* has become the language of communication. This evolution is inevitable as the use of *Bahasa Indonesia* language in national, public communication is the norm and in some cases, a necessity. This is also linked to the developing market economy, which operates alongside the existing agricultural economy. Together with an improved education system which allows the people to be more socially and economically mobile, it has become difficult to identify one's position in the social hierarchy which is based on three aspects: bloodline, wealth, and academic achievement (Anshoriy and Thoha 2005, 85). To deal with this 'difficulty' of identifying a person's social position and the accompanying linguistic norms, using *Bahasa Indonesia* is the easiest way to save one's face from addressing falsely another or using improper Javanese language: a significant faux pas.

The gradual lack of usage of *kromo* can also be related to the dissolving social hierarchy, and also to the apparent reluctance of Javanese themselves to master the language. The Javanese language is not only facing stiff competition from the national *Bahasa Indonesia* but also from the English language too. Increasingly, the younger generation considers the Javanese language as old-fashion with a complementary perception from their parents of the declining economic value of the Javanese language. More parents are sending their children to English courses as they realise the importance of English for their children's future as it is the language of the world's technology and economy. Additionally, more schools and universities are promoting new classes using English as the language of instruction, usually under the name 'International Class/Program'.[3] This situation is aggravated by the Indonesianisation in the teaching of the Humanities which focuses pupils' attention on the national History of Indonesia and the political and economic significance of the national capital, Jakarta to the marginalisation of the students' own cultural environment (Mulder 2005, 47).

On the cultural landscape, seemingly fewer and fewer people are wearing the traditional Javanese traditional dress of the *kebaya* or using the emblematic Indonesian *batik* print cloths.[4] Those who usually wear them are senior citizen traders at traditional markets. The use of this traditional dress or cloth is also seemingly limited to graduation ceremonies, Kartini Day commemorations,[5] or wedding ceremonies. The *kebaya* and *batik* are gradually being displaced by jeans

---

3   There are at least three high schools and two universities with International Class/ Programs in Yogyakarta: State High School 1, 3 and 8, and the Muhammadiyah University of Yogyakarta, and the Islamic University of Indonesia.

4   The *kebaya*, popular throughout Southeast Asia, is a fairly close fitting dress of a blouse and a sarong reaching to the ankles. Batik can refer both to the dye-resistant technique process of a fabric to the finished product itself and is emblematic of Indonesian-Malay cultural and political identity.

5   This day is celebrated as the birth day of a national female emancipation heroine, Raden Ayu Kartini (1879–1904) On this day, pupils usually wear traditional dress to schools and female workers wear the *kebaya* to work.

and t-shirts. As with the traditional art performances, the traditional dresses are increasingly reserved for special occasions and in a way a 'show' rather than an ordinary, daily expression of Javanese culture.

A significant parallel development to this phenomenon is another 'invasion' or challenge: Islamic movements with a Middle-Eastern orientation whose objectives, among others, include trying to 'purify' Islamic teaching which has been deemed to be 'contaminated' by non-Islamic values of Javanese culture, while simultaneously resisting the encroaching influence of western culture. Without a doubt, it is a common sight in Yogyakarta today to see women wearing the jilbab [head scarf] than it was several decades ago. Personal attire is probably the easiest way to mark one's identity and wearing the Muslim form of dress can be interpreted as highlighting prominently one's religious rather than ethnic identity. Much can be construed from this evolutionary assertion of identity where apparently the religious identity has assumed ascendancy and prominence over the ethnic identity. This however is not necessarily a signal of incompatibility between the Javanese and the Muslim components but could be traced to the return to spiritual values for strength, comfort and rudder in a consumer-driven world. Not an expression of religious extremism as some might be quick to conclude or retreat to impractical isolationism, a sense of cultural disenchantment with globalisation is a global phenomenon and in this sense, Yogyakarta is reflecting a worldwide expression.

On Engaging the Javanese Community

A shift in the social relations of Javanese people in Yogyakarta may be the most prominent effect of globalisation. One of the alleged benefits of globalisation included the spreading of 'democracy' at political and economic levels facilitated through the mass media and educational institutions. These outcomes have contributed to the erosion or evolution of the social hierarchy within Javanese society. This can be seen, for instance, from the number of non-royal family members holding office in government institutions or legislative bodies.

Horizontally, social relations within Javanese society has started being characterised with so-called modern society traits: rationalism, individualism, functionalism, specificity, and emotional avoidance (Waters 1995, 16); transforming the society which are often associated with other traits (Mulder 2005, 2006) like collectivity, the willingness and capacity to accept (*nrima*), mandatory good forms and etiquette (*tata karma, unggah-ungguh*), harmony (*rukun*), and mutual help (*gotong-royong*). An example of this social transformation is that more Javanese are inclined to pay workers to provide them with the services needed than to ask the help of relatives or neighbours. This is done partly to evade the social and moral obligation of reciprocity and thus facilitating both the formality and informality of relations between Javanese populations. Seemingly, fewer people are willing to help others for free, since the assistance is now perceived as 'service'. Ultimately,

Figure 7.1   Shoppers at Yogyakarta shopping mall

it is actually the same as it was before: Javanese helping each other. The difference now is that people measure their assistance in monetary terms.[6]

At a less personal level, the impact of globalisation on the community can be seen in the coming of Carrefour to Yogyakarta in 2005 (see Figures 7.1 and 7.2). This raised numerous debates on whether this giant French-origin supermarket would put traditional markets in jeopardy of losing their customers. This concern, as similarly voiced in other parts of the world on the devastating impact of global supermarket chains on local communities, has been characterised as exaggerated or alarmist as currently both Carrefour (and others supermarkets such as Makro, Hero and Indogrosir) and traditional markets still exist simultaneously. However, the impact of these global supermarket chains on the local businesses and community needs to be examined further for consequent economic, cultural and social impact.

Long lines in Carrefour check-out counters do indicate a change in Yogyakarta social life. This phenomenon does not only reflect a shopping preference, a changing

---

6   Mulder observation indicates that younger Javanese feel more comfortable having a businesslike relationship rather than a mutual contract as this would make them enjoy more independence from mutual obligation to repay such kindness to the neighbours (2005, 176–81).

Figures 7.2   Shoppers at Carrefour supermarket on a Sunday afternoon

lifestyle or even how consumer culture has permeated Yogyakarta though these are significant reasons and are interconnected. This shopping habit also underlines how social relations are being gradually revised with related outcomes. Aside from standardising products, this McDonaldisation[7] way of life at the same time rationalises social relationship: making it less personal with buyers having less time or chance to interact with sellers in the spontaneous and personally connected experience of bargaining and bantering for the best deal that a traditional market transaction can surface. Unlike the supermarket line, this socialisation process can build a relationship that in the long-run underlines the identity of a community.

On Sustaining Javanese Culture

For a culture to exist, adaptation with its ever-changing environment is inevitable. One of the ways is 'glocalisation' which involves blending, mixing adapting of two or more processes one of which must be local (Khondker 2004, 4). Javanese

---

7   'McDonaldization' encompasses the following principles: efficiency, calculability, predictability, and control of human beings by the use of material technology (See Ritzer 1993)

culture is not unfamiliar with this concept. What is known as Javanese culture today has experienced the process of blending and mixing of animism, dynamism, Hinduism, Buddhism and Islam, connoting that Javanese culture is an adaptive culture. One example is the case of the *kebaya*, which is now modified and worn with jeans instead of batik fabric, mixing the meticulously fine Javanese art with practicality value. As a response to growing number of women wearing the *jilbab*, the *kebaya* has been modified to cover most parts of the body.[8] Similarly, the Javanese traditional orchestra, the gamelan is combined with Western instruments resulting in a contemporary Javanese popular music called *campursari*.[9]

Aside from the efforts to mix and blend Javanese culture and non-Javanese ones, another effort is to try to re-discover the 'lost' values of one's culture or tradition: a form of culture fundamentalism (Waters 1995, 4). The 4th Javanese Language Congress in 2006 affirming Javanese language as a compulsory subject from elementary school to junior high school in three provinces, Central Java, East Java, and Yogyakarta is one such effort.[10] However, this political decision should also be followed with subsequent consolidation steps such as motivating and supporting the publication of Javanese literature and television programmes. Without effective follow-up programmes, this step of making Javanese compulsory might not realise the full objective of its intention in the 'rediscovering' of 'lost traditions' and 'values'.

Another effort to preserve Yogyakarta cultural uniqueness is a new bill for the acknowledgement of Yogyakarta's distinctiveness, the *Rancangan Undang-Undang Keistimewaan DIY* [Yogyakarta Special Region Distinctiveness Bill] with its draft currently being finalised and waiting further process in House of Representatives, Jakarta. The bill as drafted by Governance Department, Faculty of Social and Political Sciences, Gadjah Mada University states that the Sultan, together with Prince Pakualaman,[11] is a unified political entity forming a separate institution named *Parardhya Keistimewaan*[12] [Very Valuable Distinctiveness] and is positioned above the Governor of Yogyakarta. *Parardhya* holds the power to direct policy guidelines and a veto over four policy areas: culture, land ownership, politics and governance, and urban planning. This institution also holds the right to refuse a governor and vice governor candidate, although it has to provide a

---

8   The original *kebaya* can be too tight and transparent, with sleeveless undershirt to cover woman's chest area.

9   <http://en.wikipedia.org/wiki/Central_Java> accessed 26 July 2007.

10   <http://www.kompas.com/kompas-cetak/0609/15/jateng/41596.htm> accessed 27 November 2007.

11   *Pakualaman* is a small hereditary principality within the Sultanate of Yogyakarta. It was created in 1812 when Natakusuma (later Paku Alam I) was rewarded by helping the British quell the conflict in Yogyakarta in June 1812. <http://en.wikipedia.org/wiki/Pakualaman> accessed 27 November 2007.

12   Ancient Javanese-Indonesian Dictionary states 'Parardhya' as 'having the highest value'. See Zoetmulder and Robson 2004, 771.

strong justification for the decision.[13] The overall arrangement, at one level will give a chance for procedural democracy to take place and allow ordinary citizens opportunities to head the province. Alternatively, it can also be seen as an effort by the Sultanate to use 'democracy' as a disguise to retain vestiges of its power constitutionally, given the reality that this royal institution is slowly losing its influence. In this, the democratic component can be perceived to be a superficial effort at a democratisation process as the elected governor has to follow *Parardhya's* direction.

Essentially, this attempt could also be interpreted as negotiating the complex and increasingly urgent issues of globalising change on Javanese politics, economy, culture and of course history. In this, measured steps to retain traditions and embrace changes are practically moderated to the indigenous context: a change that does not always satisfy all sections of the Yogyakarta's population and does not always challenge the entrenched political-social power structure.

Sekaten: Finding the Middle Ground

However, a shift in social relations and the nature of collaboration within Javanese culture and market values can be best pictured in a cultural event of *Sekaten*. This cultural tradition is still the most popular Javanese ceremony in Yogyakarta. Introduced for the first time by the early Muslim missionaries in Demak Kingdom, Java in the 15th century, this ritual was then adapted by the first Sultan of Yogyakarta. Initially, this ritual was used by Islamic leaders as a means to spread the message of Islam and convert the previously Hindu and Buddhist Javanese to Islam. Those who wanted to watch the ritual performance had to declare the *Syahadat* [testimony of faith], and so denoted their conversion to Islam. The most significant part of the ritual was the *Grebeg* event. At *Grebeg,* the Sultan would make gifts of crops to his people as huge crowds gathered and competed for these bounties. The belief was that they would be granted good fortune when they kept the crops given by the Sultan.[14]

Over the years, *Sekaten* and *Grebeg* have experienced shifts in its significance and functions. In the past, on *Grebeg* day, *bupati* [district leaders] from all regions of Yogyakarta gathered to pay *upeti* [tribute] to the Sultan (Soemardjan 1982, 33). As the modern government administration developed, the committee for this event was not chaired by the Islamic leaders. Instead, it was chaired by the municipal government, in collaboration with the Sultanate. The location was also moved from the Grand Mosque to the *alun-alun* [the town square], marking the growing

---

13   Anon. (2007), 'Tata Pemerintahan Parardhya Keistimewaan Mirip MRP', *Kompas,* 6 October.

14   <http://www.kompas.com/kompas-cetak/0603/11/jogja/21918.htm>, accessed 20 June 2007 and <http://kompas.com/kompas-cetak/0304/23/jateng/274566.htm>, accessed 20 June 2007.

disassociation from the religious environment of the event. *Sekaten's* added function included the *Pesta Rakyat* [People's Festival]. This ritual was transformed into a Javanese cultural festival, with stalls selling traditional food and handicrafts and traditional performances like the *Wayang* [shadow puppet] and *Kethoprak* (traditional play). People came to gather, meet and greet others while enjoying the traditional performances. There was no longer the need to declare the *Syahadat* to enter *Sekaten*. Instead, a ticket was paid for as it continues to be affordable for most Javanese. Thus, a religious atmosphere has been replaced by a Javanese-mysticism atmosphere.

In 2003, as a sign of the changing times, *Sekaten* was privatised. The reason behind this step, as put by former Vice Head of Yogya Municipality, Syukri Fadholi, was to refresh the event, make it less monotonous. By doing so, it was expected that *Sekaten* could go nationally or even globally.[15] Privatisation, as one of the hallmarks of relentless and ruthless globalisation process, had arrived into the heart of Javanese culture. Emblematic of a corporate driven agenda of globalisation itself, privatisation impacted in a revolutionary consequence on the event. First, the name was changed into '*Jogja Expo Sekaten*': the term 'Expo' indicating just how far removed the event had come from its original intention and context. Big air-conditioned tents were erected covering the palace's front yard. This decision prompted much criticism as *Sekaten* was deemed to have been uncomfortably commercialised. As a result, starting in 2006, the Committee of the ceremony was reverted to the municipal government, after significant criticisms regarding this 'commercialisation of traditional ceremony'. Some changes have taken place since then, for instance, the entrance ticket fee has been lowered from the initial 4,000 to 1,500 Indonesian Rupiah per person. Nevertheless, the lack of religious atmosphere remains, together with the lack of Javanese cultural atmosphere. One prominent example: the tenants do not sell traditional food and handicrafts anymore. Instead, this has been replaced by clothes and hand phones, pirated compact discs, cigarettes and even motorbikes. Traditional art performances are still performed, but losing their audience.

Sekaten has indeed changed into a regular exhibition or an 'expo' although the main parts of the ceremony such as the narration of Prophet Muhammad's life and history, the performance of the palace's sacred *gamelan* and the *Grebeg* ceremony are still observed. This once sacred ceremony has turned into a mere tourism and consumerism affair with almost minimal Javanese and Islamic atmosphere. People come to the ceremony for recreational reasons, not for a cultural experience (*Kompas*, 12 April 2006). They would come simply to shop, and go home: no significant difference from what they would do at the supermarkets or at other expositions. Globalisation brings different effects to different people and can be perceived differently. For some Yogyanese, it may not matter how *Sekaten* will be as long as it is there. It is still a means of keeping the contact between the Javanese

---

15   <http://kompas.com/kompas-cetak/0402/24/jateng/873187.htm>   accessed   26 November 2007.

people and its culture, history and traditions but through consumerism – not the best or ideal way – but still having a role to play. Another set of opinion would go further to claim that the celebration of form should not be divorced from the spiritual and historical values and this must be recovered, retained and promoted.

Sekaten can be seen as representative of the multiple forces affecting this particular Javanese city. It can claim to be the focus of thrusting alien globalisation, proud Javanese culture, reviving Islamic influence or simply an annual night festival serving as entertainment centre for the people. Crucially as an evolving event, it is attempting to remain relevant and therefore adapting but the decision as to what is important, relevant and worthy for retention remains profoundly political with competing claims. In this, *Sekatan* raises more questions than answers as it negotiates the complexities and challenges of coping with globalising forces. Ultimately, the social and political dynamics in Yogyakarta will determine how *Sekaten* will be expressed in the future.

In the End ...

Ultimately, culture is not merely about exterior manifestations and declarative statements. It also encompasses deeply-held values and beliefs behind the symbolic or overt actions. Mulder argues that Javanism encompasses system of knowledge, and 'the mere performance of ceremonies in a 'traditional' manner should never be called *kejawen* or 'deeply Javanese'[16] (Mulder 2005, 42).

Notwithstanding Mulder's reminder, a Javanese city and its behavioural patterns are evolving as some of its broad physical manifestations remained. There are seemingly insurmountable challenges to resist or avoid globalisation but it can be can be negotiated, managed and adapted. Changes have come to Yogyakarta, and more will come eventually. Some changes are necessary with openness, egalitarianism, and democracy as the most prominent manifestations of change that some alleged globalisation can help bring forth. Globalisation does have a homogenising effect as it promotes the standardisation of many aspects of life: what food to eat, what kinds of music to listen, or what to put on your face. Nevertheless, 'it also pluralizes the world by recognising the value of cultural niches and abilities' (Waters 1995, 136). Friedman noted that 'dissolution of essence-appearance relation between position and person implies that there are no longer any inherent rights to rule' (Friedman 1994, 217–18).

Modern life, as surfaced by the impact of globalisation, can bring hope for some groups of ordinary Javanese in that they too have the chance to live a better life than they currently have. Undoubtedly, Globalisation is an increasingly contentious subject with myriad perspectives on its value to the global community but a personal response is the role it played in raising awareness on the importance,

---

16   In general, *kejawen* refers to the culture of the Javanese heartland that centres on the courts of Surakarta (Central Java) and Yogyakarta (Mulder 2005).

significance, value and richness of the Javanese cultural identity. This is together with the anxiety on the future role and place of Javanese cultural identity within a changing Javanese city and an increasingly homogenised world.

References

Anshory, H.M. Nasruddin and Zainal Arifin Thoha (2005), *Berguru Pada Jogja: Demokrasi dan Kearifan Kultural* (Learning from Jogja: Democracy and Cultural Wisdom) (Yogyakarta: KUTUB, in collaboration with SKH Kedaulatan Rakyat).

Friedman, Jonathan (1995), *Cultural Identity and Global Process* (London, California, and New Delhi: SAGE Publications Ltd).

Giddens, Anthony (1995), The Consequences of Modernity, (Cambridge: Polity.

Giddens, Anthony (2001), *Runaway World: Bagaimana Globalisasi Merombak Kehidupan Kita*, Translated by Andry Kristiawan S. and Yustina Koen S. (Jakarta: PT. Gramedia Pustaka Utama).

Mauzy, Diane K. and Milne, R.S. (2002), *Singapore Politics Under the People's Action Party* (London: Routledge).

Mulder, Niels (2005), *Inside Indonesian Society: Cultural Change in Java* (Yogyakarta: Kanisius Publishing House).

Mulder, Niels (2006), *Doing Java: An Anthropological Detective Story* (Yogyakarta: Kanisius Publishing House).

Ritzer, G. (1993), *The McDonaldization of Society* (Thousand Oaks: Pine Forge).

Smith, Anthony D. (1990), 'Towards a Global Culture?' in Featherstone, Mike (ed.), *Global Culture: Nationalism, Globalization and Modernity*, (London, California, and New Delhi: SAGE Publications Ltd).

Soemardjan, Selo (1982), *Perubahan Sosial di Yogyakarta* (Social Changes in Yogyakarta) (Yogyakarta: Gadjah Mada University Press).

Waters, Malcolm (1995), *Globalization* (London: Routledge).

Zoetmulder, P.J. and Robson, S.O. (2004), *Kamus Jawa Kuno* Indonesia (Ancient Javanese-Indonesian Dictionary) (Jakarta: Gramedia).

Newspaper References

Anon. (2006), 'Indikator 'Kompas', Sekaten, Antara Tradisi dan Pesta Rakyat', *Kompas Jogja*, 11 March.

Anon. (2007), 'Tata Pemerintahan Parardhya Keistimewaan Mirip MRP', Kompas, 6 October.

Internet-based References

'Central Java', *Wikipedia* <http://en.wikipedia.org/wiki/Central_Java> (last modified 23 July 2007), Wikimedia Foundation Inc., accessed 26 July 2007.

'Indikator "Kompas": Sekaten, Antara Tradisi dan Pesta Rakyat', *Kompas Cyber Media* [website] <http://www.kompas.com/kompas-cetak/0603/11/jogja/21918.htm>, accessed 20 June 2007.

'Investasi untuk Jogyakarta Expo Sekaten Rp 3 Miliar', *Kompas Cyber Media* [website] <http://kompas.com/kompas-cetak/0402/24/jateng/873187.htm>, accessed 26 November 2007.

Khondker, Habibul Haque, 'Glocalization as Globalization: Evolution of a Sociological Concept', <http://www.mukto-mona.com/Articles/habibul_haque/Globalization.pdf> (home page), accessed 15 July 2007.

'Pakualaman', *Wikipedia* <http://en.wikipedia.org/wiki/Pakualaman> (last modified 24 May 2007), Wikimedia Foundation Inc., accessed 27 November 2007.

'Pemerintah Kurang Peduli Bahasa Jawa', *Kompas Cyber Media* [website] <http://www.kompas.com/kompas-cetak/0609/15/jateng/41596.htm>, accessed 27 November 2007.

'Sekaten Pasar Rakyat yang Semakin Modern', *Kompas Cyber Media* [website] <http://www.kompas.com/kompas-cetak/0404/25/foto/988786.htm>, accessed 26 November 2007.

'Sekaten, Rekreasi Budaya di Tengah Modernisasi', *Kompas Cyber Media* [website] <http://kompas.com/kompas-cetak/0304/23/jateng/274566.htm>, accessed 20 June 2007.

'Sekaten Tahun Depan dengan Format Baru', *Kompas Cyber Media* [website] <http://www.kompas.com/kompas-cetak/0703/27/jateng/50607.htm>, accessed 26 November 2007.

'Sri Sultan Hamengkubowono X', *Wikipedia* <http://en.wikipedia.org/wiki/Sri_Sultan_Hamengkubuwono_X> (last modified 25 December 2006), Wikimedia Foundation Inc., accessed 26 July 2007.

'Sultan Minta Pelaksanaan JES Segera Dievaluasi', *e-parlemenDIY* [website] <http://www.dprd-diy.go.id/index.cfm?x=berita&id_berita=17032005212154>, accessed 27 November 2007.

'Yogyakarta', *Wikipedia* <http://en.wikipedia.org/wiki/Yogyakarta> (last modified 19 November 2007), Wikimedia Foundation Inc., accessed 27 November 2007.

'Yogyakarta Sultanate', *Wikipedia* <http://en.wikipedia.org/wiki/Yogyakarta_Sultanate> (last modified 28 August 2007, Wikimedia Foundation Inc., accessed 27 November 2007.

.

Chapter 8

Re-imagining Economic Development in a Post-colonial World: Towards Laos 2020

Michael Theno

Post-colonial Laos: A Commentary

Numerous historians have documented how the post colonialism era of Lao history has not been peaceful or prosperous for a nation already devastated by decades of secret war and the infamous reputation as the most bombed nation in human history. Notable among these historians are Vatthana Pholsena's 2006 book *Post-War Laos: The Politics of Culture, History and Identity* who noted that the 30 year civil war at the end of the French colonial period and during the American secret war left lasting wounds on the sense of a Lao nationality. The Lao government continues to be challenged by the history of Laos while the government moves towards a unifying perception of what a 21st century Laos is to look like within the context of the multiplicity of ethnic minorities that comprise a substantial number of the population.

The author avers that while the Lao People's Democratic Republic (LPDR) government remains a secretive communist political regime, the government has recognised the need to open up to the world community with a more market-based economy. Yet both Grant Evans in his 2002 book *A Short History of Laos: The Land In Between* and Vatthana Pholsena's 2006 book *Post-War Laos: The Politics of Culture, History and Identity* wrote that the lack of a generational change in leadership amongst the senior government positions in the LPDR, the ongoing and unresolved legacy of ethnic Hmong resistance to the LPDR government, the numerous ethnic hill tribes each with their own language, customs, and culture, the heavy foreign indebtedness of the Lao government, issues of government corruption, and the general diaspora of young Lao professionals all exacerbate the unfortunate reality that Laos remains one of the poorest nation states as of the early 21st century.

Some progress has been made in governance however. Again Vatthana Pholsena's 2006 history book is useful in noting that the 2003 Amended Constitution opens up the definition of Lao citizen to anyone with Lao nationality. In addition:

> With the 're-traditionalizing of the Lao state the *raison d'etre* of those concerned with preserving Lao culture has lost much of its force, especially because the Lao government controls the key sacred spaces, such as That Luang and the Palace in

Luang Phrabang, to which rituals overseas can only refer. By its very nature, the
Diaspora must refer back to Laos itself, and in this respect it has become a new
and important part of long-term change inside the country (Evans 2002, 254).

Certainly the efforts of the Lao government appear to meet the criteria of three of
the five 'meta-trends' changing the world as opined by David Snyder in his 2006
article *Five Meta-Trends Changing the World* contained in the 2006 21st edition of
Global Issues 05/06: cultural modernisation, economic globalisation., and social
adaptation. Thus the theme of this chapter is the Lao government's strategy to
leave least developed nation status by 2020. Two aspects of this strategy are the
focus of this chapter's commentary on the strategy: the relocation and repatriation
of Lao ethnic groups and the use of the various tributaries of the Mekong River
within the borders of Laos to generate direct foreign investment in Laos through
the construction of large scale hydroelectric generation stations and dams.

Social Engineering: Relocation and Repatriation of Lao Ethnic Groups

The statistics demonstrate the least developed nation status of Laos. A 2001
United Nations Lao National Human Development Report obtained from the
website www.poweringprogress.org details a 10 per cent mortality rate of children
less than five years of age, while only 20 per cent of Lao children progress to
secondary school and only 20 per cent of villages has access to electricity. A third
of the population lives below the poverty line while the probability of dying before
the age of forty is 30 per cent. Forty per cent of Lao children suffer moderate to
severe malnutrition and nearly half of Lao women are illiterate while half of all
Lao lack access to clean water. Nearly three quarters of the Lao population live on
less than US$2 a day.

'The Lao nation is one of the most heavily indebted poor countries in the
world' (Lee 2003, 116). Poverty reduction from international assistance and
investment trickles along at about 1 per cent a year. In addition, the massive
policy implementation described in the next section of damming Mekong River
tributaries for hydroelectric power generation has, and is leading to, the sizeable
displacement of various ethnic groups' villages and homes of long-standing.
T. Scudder provides a cautionary caveat regarding such displacement in a critique
of the four stage theory of resettlement in use by the Lao government: planning
and recruitment; adjustment and coping; community formation and economic
development; and handing over and incorporation:

Under conditions of extreme poverty, as in the case of Laos [hydroelectric
developments], the majority may be convinced that resettlement can improve
their living standards, but they will still have to live for years with uncertainty
as to whether that will be the case and with the risk that improvement will not
occur (2005, 44).

J. Schliesinger's book *Ethnic Groups of Laos* is about the large quantity of ethnic groups that comprise the Lao population and legitimately raises issues of how such diverse groups can merge into a national identity and be well served by the Lao 2020 goal of leaving least developed nation status (2003). The variety of cultures, living arrangements (many are tribal-based and reside in mountainous areas), languages, and historical issues of conflict rather than cooperation make attempts to serve these vastly dispersed populations a significant challenge for the national government. Schliesinger's own accounts list and describe approximately ninety such ethnicities in the northern areas of Laos alone:

> The remoteness of this mountainous, rugged and isolated region is an ideal place for people to find peace and refuge from the oppressing ethnic lowland groups. Because of these geographic barriers, the infrastructure in northern Laos is quite poor. For centuries, the main means for long-distance transportation was by boats made of tree trunks, used to navigate the rivers, especially the Mekong and Ou (xv–xvi).

Lee, writing in 2003 about Lao ethnic minorities, claimed that the 'Lao PDR is an extremely diverse country. There is no majority ethnic group, making the country a truly plural land in cultural terms' (113). Lee also cited in the same writing that inadequate land reform implementation as one of the historical sources of Lao poverty. According to Lee the Lao government has developed an eight point poverty reduction policy in its attempts to leave least developed nation status by 2020. The author avers that these eight points can best be described as a relocation commitment of the Lao population into living centres that better enable the Lao government to provide the essential services to a poorer population: infrastructure development, human and services development, and economic development.

The policy of relocation of mountainous ethnic groups and the construction of new more self-sufficient villages, self-sufficiency being defined as villages developed with the eight-point policy as guidance, has not always been well received. B. Moizo (2006) provided an overall assessment of the limits of such relocation efforts when positing that:

> Very important factors are still being left unexplored in the land issues, for example the symbolic and complex set of relationships that exist between swidden farmers and the area they live in. Also the importance of the identity components in the way people perceive, get access and use the land and more specifically their territory, made up of interrelated things – the village, the fields, the fallows, the forest, the secret and sacred places ... [Mountain dwelling ethnic groups] are caught in a schizophrenic turmoil in which their former perceptions are in direct conflict with the land law. In the past, these farmers did not have titles to the land but they controlled access and managed the use of resources within their territory. Nowadays, they may have land allocated to them and some land titling, but they no longer have the full right to manage their territory the way they want (43–4).

Then there is the Hmong ethnic group. M. Theno and M. Speck, writing in 2006, described a history of a fiercely independent group that sided with, or was used by, the United States to prosecute a secret war against the Pathet Lao. When the United States abandoned its ally in 1975 the resulting Lao government's war against its ethnic group entity led to massive flight into Thailand, and thus a fragmentation of a people. 'When the fight became a losing one, they were left to shift for themselves in a country suddenly hostile. Some of them escaped' (Champeon 2004, para. four). This refugee diaspora culminated, or so most policymakers hoped, in the permanent resettlement of approximately ten thousand Hmong to the United States in 2004–2006 in what Theno and Speck identified as a moment of enactable policy not previously available or likely to be available again (2006, 366). There they joined the already estimated 100,000 Hmong previously resettled or born in their new country (Champion 2004, para. nine) (See Figure 8.1).

The Hmong dilemma continues to this day. In the 21 September 2007 article 'Hmong to be Sent Back to Laos', The *Bangkok Post* reported that approximately 8,000 Lao Hmong, comprised of those left behind in the US relocation of 2004–2006 as well as Lao Hmong whom supposedly fled their home country to join their fellow displaced Hmong in Thailand, have been 'contained' in a compound in the north central area of Thailand where they will be repatriated to Laos in 2008. The *Bangkok Post* also reported that the Lao government has so far refused to allow human rights observers to participate in the repatriation and resettlement plans. This presents itself as a human rights conundrum as well as identification in real human terms of the limits to the enactable policy moment identified by Theno and Speck (2006).

However, the Lao Hmong within the national borders continue a combative and mutually distrustful relationship with their national government. Jamaree Chiengthong's 2003 article *The Politics of Ethnicity, Indigenous Culture and Knowledge in Thailand, Vietnam and the Lao PDR* included in the edited book *Social Challenges for the Mekong Region* alleged that stereotypical negative images of the Hmong were inculcated in both media and government reporting and official dialogue. The *Bangkok Post* printed an opinion article by Thomas Fuller on 18 December 2007 that reported an unknown number of ethnic Hmong who live in Laos continued to live in hiding and still hope that the United States government will rescue them from what Fuller reports is the on-going persecution they suffer at the hands of the Lao government. This leads back to the relocation policy of the Lao government and its effects on the various ethnic groups.

Chiengthong described the development policies of resettling Lao ethnic groups as one of forced relocation into 'manageable locations' and cited McCaskill's 1997 terminology of 'domestication rather than development' to describe the Lao government policies of better serving its ethnic group citizenry (2006, 156). Chiengthong also averred that the Lao PDR governments' relocation policies focused on resettlement from highland areas deemed unsuitable for cultivation into more permanent areas for cultivation (2006, 156). However, this may result in the unintended consequences of even more ecological damage and social dislocation. Addressing the issue of social dislocation, Chiengthong averred that 'accounts

Figure 8.1   Lao-Americans celebrating New Year in Fresno, California, December 2007

of many ethnic groups revealed what one may call the "narration of inferiority" when coming into contact with lowlanders and state authorities' (159–60). Finally he states that the politics of 'indigenous knowledge' has given way to the official policy of 'inclusion and exclusion' in the name of economic development (160). The author puts forward the possibility that, as all too often occurs in the pursuit of a higher standard of living, *anomie* is the price paid by the generations directly affected by the economic development implementation.

Is there an alternative to this situation? The author puts forward that the assertion that history, particularly Lao history, suggests there is no feasible alternative. Champeon claims that ultimately, these relocated ethnic groups 'will be free only to be identified with everybody else; and the land itself shall no longer speak to its former caretakers' (2004, para. ten). The price of moving towards development status with a policy of increasing the self-sufficiency of local areas may well be the loss of distinct identities over the generations. The author posits that this is not without historical precedent in the world.

Geographical Engineering: Dams and Water as a Competitive National Advantage

To understand the Lao decision to use the tributaries of the Mekong River within Lao national borders as a source of generating direct foreign investment, it is useful to provide some understanding of the strategic significance of the Mekong River and its tributaries. Milton Osborne's 2000 article *The Strategic Significance of the Mekong* printed in an edition of the Journal *Contemporary Southeast Asia* documented several fascinating facts about the Mekong River. These include the following: unlike other large river systems of the world the Mekong has not played a unifying role in history, the Mekong flows through China, Myanmar/Burma, Laos, Thailand, Cambodia, and Vietnam, the reality that Western awareness of the Mekong has largely been based on the river's military significance in the various colonial and Cold Wars fought in post World War Two western history, modern Mekong policies are part of the debate about the appropriate uses of the Mekong as a source for economic development, and current practices centre on dam projects. In particular Osborne noted that Thailand's construction of the Pak Mun Dam on the northeastern Thai Mekong tributary, the Mun River, in the 1990s has failed to meet projected hydroelectric output while the dam and reservoir displaced up to seven times the number of persons estimated in the planning for the dam. Chris Greacen and Apsara Palettu wrote in their 2007 chapter *Electricity Sector Planning and Hydropower* in the edited book *Democratizing Water Governance in the Mekong Region* that Thai electric power officials have deferred any further plans for damming in apparent recognition of the failure to perform vis-à-vis the human displacement costs.

Again the author cites both Osborne's 2000 article and Lee's 2003 article in noting that Laos has chosen a different strategy. The Lao government policy objective of moving out of least developed nation status by 2020 is based on increased earnings of foreign exchange via sales of hydroelectricity. The Lao policy objective builds on many years of using international funding consortia to build a series of dams despite widespread criticism from international environmental organisations. Indeed Greacen and Palettu described Laos as 'the battery of Asia' in their 2007 article cited in the previous paragraph.

K. Theeravit's research appears to endorse such measures: 'We do not know how many people in the Mekong Region are impoverished because of globalisation. [This is because] it is difficult to separate the traditional impoverishment from that caused by globalisation' (2003, 70). Some conservationists have remarked that if the set of internal controls built into the development and management of the most recent dam projects, such as the Nam Theun 2 dam, are properly achieved then these projects could become a model for use world-wide (Rowse 2002). 'The Mekong River Commission holds up that the Nam Theun 2 project is an excellent example of an integrated approach to dam building' (Murphy 2007, 79).

Regardless of the Nam Theun 2 projects potential as described by Murphy in the previous paragraph the demands for Lao dams continue unabated and cross national boundaries. The *Bangkok Post* reported on 26 December 2007 that Laos is considering a mega-dam on the Mekong in response to Vietnam's energy needs so as to meet the increasing energy needs of Vietnam. 'Vietnam's main energy company expects ... a dam near Luang Prabang [the former Lao royal capital] that would dwarf existing dams in the landlocked country' (Murphy 2007, 5). Is there any limit to the feasibility of continued dam-building?

Perhaps G.U. Aliagha and K. Goh's observations about Lao government's intentions to use its water resources offers some answers with the following statement: 'The national economic development process is to be based on the wealth of natural resources, especially its hydropower resources for both local use and for foreign exchange earning' (2006, 320). A caveat to this observation is the one of Scudder (2005) who advocates large dams 'as a flawed yet still necessary development option' and cautions that the numerous flaws may cancel out advocacy when the reason for large dam development is averred:

> Large dams remain a necessary development option to deal with the needs of a human population that is expanding beyond the carrying capacity of the world's life support systems. That is the tragedy. A development strategy that over the longer term is degrading critical natural resources remains necessary, at least over the short term, in countries such as Laos ... to provide foreign exchange for development purposes (1–2).

The author observes that these writings provide ambiguous answers at best. Perhaps the limits of direct foreign investment will determine the limits of Lao dam-building projects. Perhaps unintended ecological effects will determine the limits of Lao dam-building projects. Perhaps the Lao people's popular opinion will rise up, as it did in its neighbour Thailand in the 1990s, and turn against the feasibility of future dam-building at some time in this decade. The author posits that the limits to dam-building in Laos will emerge as a combination of the above plus other unforeseen circumstances.

Concluding Observations

Hoskisson et al. provide some support for the capacity-building opportunities of multinational enterprises that finance natural resource developments such as hydroelectric dams (2006). 'It is anticipated that as markets emerge [a] convergence of economic systems is occurring in terms of legal institutions and governance mechanisms' (Hoskisson et al. 2000, 263). Laos must develop such systems, institutions, and mechanisms to move onto the next economic level in the estimation of the author.

As a least developed nation-state with dispersed and isolated ethnic minorities, any attempt to build a comprehensive social welfare system is much further in Laos' future, given the country's largely subsistence economy and despite a government focus on large scale exploitation of natural resources for economic development.

Rosenau averred that 'there is no inherent contradiction between localising and globalising tendencies' in what he described as 'globalisation's eventual predominance' (2006, 49–50). A specific example of this is the Lao government's plan to finalise membership in the World Trade Organisation (WTO) in 2008 as reported by the *Lao News Agency* on its website on 29 November 2007. As the experience of China has shown, such membership in a worldwide trade assembly such as the WTO does not guarantee human rights and continuing violations of these rights may be evident in the Lao government's social engineering policies, as described earlier.

As the sixtieth anniversary of the United Nations Universal Declaration of Human Rights approaches, the author S. Tharoor writing in the 2006 21st edition of the edited work *Global Issues 05/06* cautioned for a 'right to development' in 'universality, not uniformity' in any global definition of human rights (188). J. Diamond wrote that 'all such [policy] decisions involve gambles, because one often can't be certain that clinging to core values will be fatal, or (conversely) that abandoning them will ensure survival' (2005, 433). Diamond also states that societies that survive and flourish develop a predisposition toward a social and policy flexibility to adapt to changing conditions as well as the financial and technological capability to implement such a predisposition (434). Laos may well begin with some form of national reconciliation with the various ethnic groups who at one time in Lao history may have resisted efforts to establish the Lao People's Democratic Republic.

Arguably the author posits that one method to assist in such reconciliation is to encourage increased individual Lao contacts and interactions with foreigners through tourism. Writer Robert Davis described in the *Bangkok Post* on 29 December 2007 'the making of an eco-tourist nation' in Laos (5). Davis observed that 'this was Lao, I thought; a warm hospitality like none other I have ever witnessed around the world' (5). The author puts forward that such individual observations multiplied numerous times may well unintentionally open up Laos more than any government policy.

The author hypothesises that to develop such flexibility requires education, innovation, and intergenerational changes in governance. However, flexibility also requires a rise to a level of economic development in which increasing numbers of a nation's population can themselves rise above subsistence economics. That the Lao government is even willing to attempt such a beginning of this progression is, in itself, to be commended and supported. Commendation and support may then very well lead toward a more open Lao government and society.

References

Aliagha, G.U. and Goh K. (2006), 'Essentials for Mobilizing Regional Cooperation in the Mekong River Basin', in Goh, K. and Yongvanit, S. (eds), *Change and Development in Southeast Asia in an Era of Globalisation* (Singapore: Pearson/Prentice Hall).

Chiengthong, J. (2003), 'The Politics of Ethnicity, Indigenous Culture and Knowledge in Thailand, Vietnam and Lao PDR', in Kaosa-Ard, M. and Dore, J. (eds), *Social Challenges for the Mekong Region* (Chiang Mai, Thailand: Social Research Institute).

Davis, R. (2007), 'Welcome to Laos: The Making of an eco-tourist nation', *Bangkok Post*, 29 December.

Diamond, J. (2005), *Collapse: How Societies Choose to Fail or Succeed* (New York: Penguin).

Evans, G. (2002), *A Short History of Laos: The Land in Between* (Australia: Allen and Unwin).

Fuller, T. (2007), 'CIA's Army in Laos Still on the Run: Hmong Who Fought alongside Americans Continue to Live in Hiding, Hoping Washington Will Come to their Rescue', *Bangkok Post*, 18 December.

Greacen, C. and Palettu, A. (2007), 'Electricity Sector Planning and Hydropower', in Lebel, L. et al. (eds), *Democratizing Water Governance in the Mekong Region* (Chiang Mai: Thailand: Mekong Press).

Hoskisson, R. et al. (2000), 'Strategy in Emerging Economies', *Academy of Management Journal* 43:3, 249–67.

'Laos Mulls Mekong Mega-dam: Vietnam Wants to Build Project as Power Source', (2007), *Bangkok Post*, 26 December.

Lee, K. (2003), 'Social Challenges for the Lao PDR', in Kaosa-Ard, M. and Dore, J. (eds), *Social Challenges for the Mekong Region* (Chiang Mai, Thailand: Social Research Institute).

Moizo, B. (2006), 'Kmhmu Responses to the Land Allocation Policy: A Case Study from the Lao PDR', in: Goh, K. and Yongvanit, S. (eds), *Change and Development in Southeast Asia in an Era of Globalisation* (Singapore: Pearson/Prentice Hall).

Murphy, C. (2007), 'Slowly up the Mekong', *Far Eastern Economic Review* 170:1, 78–9.

Osborne, M. (2000), 'The Strategic Significance of the Mekong', *Contemporary Southeast Asia* 22:3, 422–9.

Pholsena, V. (2006), *Post-War Laos: The Politics of Culture, History and Identity* (Singapore, The Nordic Institute of Asian Studies; Institute of Southeast Studies Press, Silkworm Book).

Rosenau, J.N. (2006), 'The Complexities and Contradictions of Globalization', in Jackson. R.M. (ed.), *Global Issues 05/06*, 21st edn (Dubuque: Iowa, McGraw-Hill/Dushkin).

Schliesinger, J. (2003), *Ethnic Groups of Laos: Volume 4. Sino-Tibetan-speaking Peoples* (Bangkok: White Lotus).

Scudder, T. (2005), *The Future of Large Dams: Dealing with Social, Environmental, and Political Costs* (Sterling, Virginia, Earthscan).

Snyder, D.P. (2006), 'Five Meta-trends Changing the World', in Jackson. R.M. (ed.), *Global Issues 05/06*, 21st edn (Dubuque: Iowa, McGraw-Hill/Dushkin).

Tharoor, Sashi (2006), in: Jackson, R.M. (ed.), *Global Issues 05/06* 21st edn (Dubuque: Iowa, McGraw-Hill/Dushkin).

Theeravit, K. (2003), 'Relationships Within and Between the Mekong Region', in Kaosa-Ard, M. and Dore, J. (eds), *Social Challenges for the Mekong Region* (Chiang Mai, Thailand: Social Research Institute).

Theno, M. and Speck, M. (2006), 'The Lao Hmong of Wat Tham Krabok: A Moment of Enactable Policy', in Goh and Yongvanit, (eds), *Change and Development in Southeast Asia in an Era of Globalisation* (Singapore: Pearson/Prentice Hall).

Internet-based References

Champeon, K. 'The Last of the Free', *Things Asian* [website], (published online 2 August 2004) <http://www.thingsasian.com/stories-photos/3025> accessed 26 November 2007.

'Hmong to be Sent Back to Laos', 21 September 2007, *Bangkok Post* [website], (published online 21 September 2007) <http://www.bangkokpost.com/topstories/topstories.php?id=121873> accessed 26 November 2007.

'Laos Prepares to Join WTO', 29 November 2007. *Lao News Agency* [website]. (published online 29 November 2007) <http://www.Kpl.net.la/> accessed 5 December 2007.

'National Human Development Report Lao PDR 2001', *United Nations Development Programme* [website]. (published online 2001) <http://www.undplao.org/whatwedo/factsheets/humandev/nhdr%20final.pdf> accessed 27 November 2001.

Rowse, B. 'Poverty Stricken Laos Looks to Dam for Salvation', *Things Asian* [website]. (published online 2002) <http://www.thingsasian.com/stories-photos/2209/2160298/2/brt0_art;jsessionid=29AC7D8D0599FCD4D7C5AD69A94D76F2> accessed online 16 February 2007.

Chapter 9

When was Burma? Military Rules since 1962

Nancy Hudson-Rodd

Introduction

Our Three Main National Causes:

Non-disintegration of the Union
Non-disintegration of national solidarity
Perpetuation of sovereignty

People's Desire:

Oppose those relying on external elements, acting as stooges, holding negative views
Oppose those trying to jeopardize stability of the State and progress of the nation
Oppose foreign nations interfering in internal affairs of the State
Crush all internal and external destructive elements as the common enemy
(Ministry of Information, State Peace and Development Council, 2007).

The three main goals, 'National Causes' of the *Tatmadaw* (Burmese military) and the means of achieving these goals, 'People's Desire' are repeated over and over as the ruling State Peace and Development Council (SPDC) regime dictates that these words be placed on the front pages of every publication, magazine, newspaper and book (English and Burmese languages). 'People's Desire' also is writ large on billboards throughout Burma (see Figure 9.1 and following figures). 'To uphold "Our Three Main National Causes" is the bounden duty of the entire national people of the Union of Myanmar[1] ... The ultimate goal of the national people is emergence of a discipline-flourishing modern developed democratic nation, in which all the national races desire to live keeping egg and nest intact' (Ministry of Information 2007, a). The SPDC military regime, in a call of unity to

---

1 The military junta in June 1989 decreed the country's name change Myanmar for Burma and the names of most major cities (Yangon for Rangoon), states and divisions (Pegu for Bago), and geographical features (Ayeyarwady for Irrawaddy River and Division). I do not endorse the regime's interpretations or changes. In this chapter, I use former spellings for commonly used names of states and divisions, except for quotes. There are no standardised transcriptions in the Roman alphabet for Burmese, Karenni, Mon, and other ethnic languages. Words are spelled variously according to different spelling conventions.

Figure 9.1   'People's Desire' banner, Rangoon, 2008

end ethnic violence of separation, claims to be fighting enemies of the state, so as to be able to create a modern Burma, as an independent nation. The military rulers use the language of belonging, a call to home in the independent Burma. They claim to be waging war to bring peace and development. But, the very change of name from Burma to Myanmar has been viewed as an aggressive policy of 'Myanmafication'. In essence, many non-Burman leaders regard this as another act of 'Burmanisation', the latest stage in attempts to deflect and destroy minority cultures, while establishing a single identity for the country (Houtman 1999, 137–56).

Present debates concerning Burma or Myanmar centre on the promotion of quite different perspectives of the history of the country. According to General Than Shwe, the military regime rulers view a continued peaceful existence:

> A variety of national races live in the territory of our nation Myanmar, and each and every part of the nation is like a small union where different nationalities reside. We Myanmars have been living together in unison so long that we are inseparable. And love, friendship and unity are a must for all of us (2007, 3).

> Thanks to the unity and farsightedness of our forefathers, our country has existed as a united and firm Union and not as separate small nations for over 2,000 years (2002 speech to the University for Development of National Races).

Figure 9.2   Town Hall, Rangoon, 2008

These assertions are disputed by many authors and international organisations and evidenced by the vast numbers of asylum seekers fleeing Burma over the past decades under military rule. Scorched earth campaigns against insurgents have resulted in hundreds of thousands of refugees fleeing to Bangladesh, India, Malaysia and Thailand.

In this chapter, I explore the complexities of nation in Burma and the central problem of military rule for over four decades of violence committed by the regime on the citizens of Burma[2] with civilian populations suffering the most.

Burma

Four major river systems feed Burma's rich fertile land. The country is endowed with natural resources of arable land, teak and other hardwood forests, minerals (natural gas, oil, petroleum, lead, silver, tin, zinc), precious and semi-precious

---

2   The term Burman is usually used in English for the majority ethnic group and Burmese for language, citizenship or other national terms. For example, a member of Shan ethnic group is called a Burmese citizen. Since 1989, the military regime has used *Bamar* and *Myanmarese* for Burmese.

gems (jade, rubies and sapphires). Burma's plentiful natural and human resources are sufficient to sustain the present population of 54 million and future generations while maintaining its diverse ecological milieu. Yet, Burma is a most impoverished and isolated nation with per capita national income level below that of its neighbours: Bangladesh, Cambodia, and Laos. Burma's impoverishment results from decades of repressive policies implemented by successive military regimes since 1962. These regimes are responsible for a mismanaged economy (Burma Economic Watch 2006), extreme poverty with most families spending 70 per cent income on food compared to the global benchmark of 50 per cent as an indicator of poverty (The Economist 2004; Food and Agriculture (FAO) and World Bank 2004, 40), gross human rights abuses (AAPP-Burma 2005; Amnesty International 2007; Human Rights Watch 2007), continued use of forced labour, (ILO 2006) human trafficking (US Department of State 2007), and regime attacks on ethnic minority groups (KWO 2007) in border areas.

Burma's isolation is due to violent actions committed by the regime in 1988. About 3,000 citizens taking part in countrywide peaceful demonstrations were killed by the military which seized control, suspended the constitution and declared martial law. In 1990, multi-party democratically held elections resulted in citizens of Burma overwhelmingly voting out the regime in favour of the National League for Democracy, led by Daw Aung San Suu Kyi and U Tin Oo, who both remain under house arrest. The regime refused to implement the election results rather harassed, detained, imprisoned or forced elected members into exile (Silverstein 1996a). A decade later (April 2000), at least 40 elected members of parliament (MPs) remained in prison, held under harsh conditions which reportedly included 'cruel disciplinary practices and torture, lack of proper medical care and insufficient food', contributing to the deaths of three members in prison and three shortly after release (Inter-Parliamentary Union 2000, 3). Almost two decades after the elections, thirteen MPs remain in prison (NCGUB 2007, 2). Other elected members fled to Manerplaw, headquarters of the Karen on the Thailand/Burma border. With the support of 'the Democratic Alliance of Burma, a political front of ethnic minority and Burman groups, formed the National Coalition Government of the Union of Burma (NCGUB) as a rival to SLORC' (State Law and Order Restoration Council). Dr Sein Win, the leader, is Aung San Suu Kyi's cousin (Silverstein 1996b, 1). The NCGUB members lobby at the United Nations, speak often to parliaments, political leaders, and the press as the credible representatives of Burma. Delegates of the 116th Assembly of Inter-Parliamentary Union (IPU) held in Nusa Dua, Bali, Indonesia (30 April 2007) as delegates acknowledged that 'misrule' of Burma's regime had negative affects on the region. The IPU urged the military authorities in Burma to release all 'MPS-elect still in detention immediately and unconditionally, and to guarantee their physical integrity'. ASEAN Inter-Parliamentary Myanmar Caucus (AIPMC) called for unconditional release of all political prisoners, including Aung San Suu Kyi, ethnic leaders, and elected members of parliament (NCGUB 2007, 1).

Burma has become increasingly isolated from the World Bank, the International Monetary Fund, and the Asian Development Bank. The World Bank has not approved any new loans to Burma since 1987 and has no plans to resume its programmes given continued rule by military regime. Burma is currently in arrears to the World Bank and has not enacted essential economic and other reforms. The Asian Development Bank and the International Monetary Fund have not made new loans to Burma since the 1980s (US GAO 2007, 7).

The International Monetary Fund's report on Burma (September 2006) estimated the per capita gross domestic product (GDP) of Burma in 2005 to be US$ 170, less than half of the per capita GDP of the poorest countries in the region. For example, per capita GDP in Cambodia for the same year is estimated to be US$ 350, for Laos the figure is US$ 400, and for Bangladesh, US$ 440. Agriculture accounts for more than half of the Burma's GDP and employs over two thirds of all labour. Many of the estimated 19 million farmers increasingly struggle to provide a living for their families. The rural poor (about 70 per cent of the poor) face a range of problems: limited access to land, credit, paid employment, education and health facilities; and a dwindling supply of natural resources in the context of increasing environmental degradation (US CIA 2006). Current estimates are that 30 per cent of rural people are landless, while some parts of the country experience much higher rates. The proportion of households with medium size farms (5–10 acres) in 2003 was estimated to be about 15 per cent and a further 37 per cent of rural households depend on small or marginal farms with less than five acres. Only 7.6 per cent of households farm more than ten acres of land (FAO and World Bank 2004).

The UN Development Program (UNDP) ranked Burma 132 out of 177 countries in its 2007/8 annual human development index based on social and economic indicators (life expectancy, educational attainment, and adjusted real income). According to the World Health Organisation (2000), Burma's health care system is second worst in the world, 191 out of 192 nations as health indicators show that high mortality and morbidity rates are due to preventable diseases. Only Sierra Leone ranked lower (Council on Foreign Relations 2003). Rural populations' increased vulnerability is due to a burden of preventable diseases, HIV/AIDS, tuberculosis and malaria, and diarrhoea which is the major killer of children. In 2005, Burma accounted for over half of all malaria deaths in Asia (UN, Second Regular Session 2006, 4). Burma contrasts with other countries in the region in spending more money on defence than on health and education combined (Selth 2002, 135). Increasing problems with HIV/AIDS and narcotics abuse in Burma attract international attention (Garrett 2005, 37), but there is a dearth of accurate field research and data. Health needs of Burma cannot be addressed in face of denial of access to information, no freedom of academic research, and no ability to disseminate research findings. Secrecy and censorship severely impact the health of people in Burma.

Based on civilian testimonies from ten out of 14 states and divisions, food scarcity was shown to directly result from militarisation and its accompanying lack of freedom of expression (Asian Human Rights Commission). The United

Nations estimates that the Burmese population spend up to 70 per cent of their monthly income on food (Kyi May Kaung 2007). As the military regime expands into non-Burman ethnically dominated border areas in Arakan and Chin States of western Burma and Mon, Karen, Shan, and Karenni States of eastern Burma, poverty becomes more acute and humanitarian needs are especially high after years of conflict and instability. Burma facing severe economic stagnation, widespread poverty and serious health and social welfare issues remains one of the 'countries of most concern' in the Asia Pacific as it falls ever further behind in attaining their millennium development goals (Asian Development Bank 2006). The majority of the population are poor, the land is increasingly impoverished and most people survive day to day, in fear of the military regime (Hudson-Rodd and Myo Nyunt 2005). Those living in rural areas are most severely affected in all measures of health, education, transportation, and social services.

> The relentless attempts at totalitarian regimes to prevent free thought and new ideas and the persistent assertion of their own rightness bring on them an intellectual stasis which they project on to the nation at large (Aung San Suu Kyi 1995, 175).

Burma faces a problem not of a lack of financial resources, but rather a regime which denies freedom of expression, censors all material, and gives no space to the views, ideas, and contributions of the dis-interested educated, knowledgeable, and learned people of Burma. There are strong reasons to believe that freedom of expression makes possible other rights to be fully realised (Lasner 2005, 250). An open and independent media can promote not just democracy and human rights but also economic development in whose name freedom of the press is often sacrificed. The lack of media freedoms facilitates corruption which increases the burden of poverty and prevents individuals from making any informed decisions. Amartya Sen (1999a, 15–17) argues that development consists of the removing of a variety of 'un-freedoms' that leave people with restricted options to explore and create their lives. Restrictions on media reinforce the corruptions in the process of Burma's developmental policy-making. As such the denial of freedom of expression is central not just to the regime's continued power but also to the lack of economic development in Burma.

Economic planning proceeds in Burma with no public input, reliable economic data, or official accountability. Without freedom of academic research and the ability to disseminate research findings, there can be no public debate informed by independent sources. The *Review of the Financial, Economic and Social Conditions* published under different names by the Ministry of Planning and Finance since military rule in 1962 was publicly available until 1998. This had been the main source of annual data on the Burmese economy as projected by the regime itself. Since 1998, the public has been denied access to even this document. While the accuracy of the data in such publications could never be fully ascertained, the publication did provide a record for comparison and evaluation by researchers.

Figure 9.3   Road near Beautyland, 2008

The regime also ceased publishing annual budget estimates in 2001/2. The Central Statistical Organisation (CSO) releases erratically the *Selected Monthly Economic Indicators*. The CSO claimed to have completed a *Household Income and Expenditure Survey 2000*, but the report was never released. The regime restricts access to official documents and reports conducted by international organisations.

In-country researchers are not able to use information from publications available to those outside Burma. For example, the International Monetary Fund (IMF) makes regular reports on Burma, but the regime denies release of these reports within Burma. As one economist in Burma points out most cogently, 'as well known to all responsible members of the Burmese mass media, both within the country and abroad, academics like us play no role and are completely out of the picture in the decision-making process regarding issues that are of major concern to the people of Myanmar' (Maung Myint 2007).

Figure 9.4   Street gamers, Rangoon, 2008

By all measures, people of non-Burman ethnic minority groups living in the border regions of Burma are more profoundly affected by the activities of the military regime.

Ethnicity in Burma: History and Management

Ethnic minorities make up about one-third of the population and occupy almost one half of the land. The seven states and seven divisions reflect these ethnic groups. The Union of Burma comprised Kachin State, Shan State, Karenni State,

Figure 9.5   Shwedagon Market, November 2007

Figure 9.6   Street phone, Rangoon, 2008

Chin Special Division and Burma proper at the time of independence on 4 January 1948. Karenni State was renamed Kayah State and Karen State formed in 1951. According to the Constitution of the Socialist Republic of the Union of Burma (Myanmar), promulgated on 3 January, 1974, there were seven States and seven Divisions. At present the Union consists of the same seven States and seven Divisions (Hla Tun Aung 2003, 604). The country is currently divided into seven primarily Burman ethnic Divisions (*tain*): Irrawaddy, Pegu, Magwe, Mandalay, Rangoon, Sagaing, and Tenassarim and seven ethnic States (*pyi nay*): Chin, Kachin, Karen, Karenni, Mon, Arakan, and Shan.

Burma is a multi-cultural, multi-religious, multi-linguistic nation with officially 135 national 'races' with the eight 'main stocks' being 'Kachin, Kayah, Kayin, Chin, Bamar, Mon, Rakhine and Shan' (Ministry of Information 2007, 4). Hla Tun Aung, former rector and former professor of Geography at University of Mawlamyine using the 1983 Census conducted by the military regime, describes ten main groups in Burma with the majority, about 69 per cent of the country's residents Burman and the Shan at 8.5 per cent, being the largest ethnic minority group. The eight other groups include: the Karen (6.2 per cent), Arakan (4.5 per cent), Mon (2.4 per cent), Chin (2.2 per cent), Kachin (1.4 per cent), and the Karenni (0.4 per cent), other indigenous groups (0.1 per cent), and (5.3 per cent) mixed Myanmar and foreign groups (Hla Tun Aung 2003, 224). Each of the major groups is officially sub-divided into smaller minority groups. For example the 135 sub-nationalities comprised: 12 for Kachin: 9 for Karenni; 11 for Kare; 53 for Chin; 9 for Burman; One for Mon; 7 for Arakan: and 33 for Shan (*Lokethar Pyithu Neizen*, 26 September 1990).

These figures however are controversial outside Burma as there are no reliable statistics on population or ethnic distribution in Burma as the last comprehensive population census was carried out by the British in 1931 (Frontier Areas Committee of Enquiry 1947). The regime and the armed opposition groups have each made estimates of population with ethnic activists claiming larger numbers for each group other than Burman. These figures are all influenced heavily by political and other considerations. Hla Tun Aung (2003, 224) addresses the 'differences of opinion regarding the inclusion of some sub-groups under some main groups' and not others, by not arranging ethnic groups according to linguistic divisions, but rather by the territories wherein the majority of each sub-group reside. This geographical way of studying ethnicity illustrates the diverse mix of ethnic peoples within each state and division of Burma. For example, in Karenni and Arakan States, the majority of the population were of Karenni and Arakan ethnicity respectively. These were exceptions as other states and divisions had a mix of groups intermingled.

Far from a unitary state, Burma since independence in 1948 has been riven with complex ethno-territorial conflicts between the central government and ethnic groups seeking autonomy. The state has never exercised complete administrative, judicial, or executive power throughout the border areas. Historically there never existed a unified nation state of Burma. Rather there were 'state spaces' and 'non-state spaces'. According to Kahing Soe Naing, general secretary of the National Democratic Front (NDF):

> The British imperialists integrated all the countries, states and areas of the people of Karen, Karenni, Mon, Burma, Arakanese, Kachin, Chin, Shan, Pa-O, Palaung, Wa, Lahua ... etc., some were forcibly occupied, some were agreed on as protected areas and were formed into the so-called Burma. From then on, Burma came into existence (2000, 15).

At independence, great differences existed between the legal status of the new government's state power over Burma and the control that it actually exercised over people and place. But ethnicity was not the only dimension of change. Myint Thant (2001) has identified how the British left many minority societies untouched, but abolished the traditional institutions of power among the Burma majority thereby undermining any modernising influences which may have arisen from the royal court system. Only the Buddhist *sangha* (clergy) were left to any extent uninterrupted to the present. This power vacuum was filled by the Burmese military *Tatmadaw*. The other legacy of British colonialism formed in the struggle for independence was one based on a form of Burmese ethnic nationalism of past 'memories' rather than one of 'a newer identity which would incorporate the diverse peoples inhabiting the modern state' (Myint Thant 2001, 254). The rarely challenged view of independence struggle was that a successful war had been fought against colonial rulers by Burman nationalists who then only needed to unite all the other indigenous groups to restore a historic Burma. This Burma never existed. The dilemma of modern Burma was eloquently explained by Professor Tun Aung Chain, a Karen peace negotiator and vice-chair of the Myanmar Historical Research Commission in Rangoon in 2000. He suggested that 'Myanmar nation-building' was a political priority since independence and argued that a 'type of history' had been developed which projected modern 'political aspirations' into the past. In the process, insufficient account was taken of other 'equally valid political and cultural centres' such as the Mon, Rakhine, and Shan States. But since the 'creation of Myanmar nationhood out of its ethnic diversity still remains on the agenda', Tun Aung Chan (July 2000, 11) urged that 'the formulation of a more sophisticated history' remained a big 'challenge' to the country's historians.

When Britain granted independence in 1948, there was a great difference between the legal status of the new government's state power over Burma and the control that it actually exercised over people and place. Less than 12 weeks after independence, the Communist Party of Burma led an armed rebellion against the government. Other communists, irregular forces, ex-Patriot Burmese fighters, disaffected Muslims in Arakan (Rohingyas) joined with Karen, Shan, Padung, Pao, Mon, and Kachin in the struggle against the new independent government of Burma. Insurgent ethnic groups controlled almost 50 per cent of the country. Over six decades later, the military junta in 2007 seeks control of a unified state, while several ethnic groups seek control of their own 'nations'. History is being written from a variety of perspectives.

Institutionalised oppression, ethnic fragmentation, and political distrust have all featured in Burma since independence from the British in 1948. As noted earlier, historically there never existed a unified national state of Burma but rather a recognition of 'state spaces' and 'non-state spaces'. In the former the subject population were settled densely into permanent communities, in the heartland of Burma, producing a surplus of rice, easily appropriated by the state. In the latter, the sparse, highly mobile populations were scattered widely, practicing swidden forms of cultivation and a mixed economy, which reduced the possibilities of state appropriation. More groups (Karen, Shan, Padaung, Pao, Mon, Kachin) left the government forces and joined the struggle against government control in response to attacks by Burma Union's paramilitary police force. As noted earlier, almost 50 per cent of the country was controlled by insurgent groups (Smith 1999; Lintner 1989). These armed insurgencies weakened the U Nu democratic government and a caretaker military government took control. General Ne Win seized power in 1962, dismissed parliament and abolished the Constitution (Callahan 2003).

The military has ruled Burma continuously since 1962 dominated until 1988 by General Ne Win and the military party, the Burma Socialist Program Party (BSPP). The military implemented socialist rules while political power resided in the Revolutionary Council. State power was centralised and 'suspected' ethnic minority leaders were imprisoned (Silverstein 1977, 96). In 1974 a new military constitution was written which declared a unitary state and denied any autonomy to the states. Seven divisions and seven states with arbitrary boundaries were created with centralised administration.

There were no free elections, freedom of expression and association was denied. Resistance to the regime was quelled. Physical coercion of political prisoners was common. Ethnic insurgencies intensified in response to increased military power (Silverstein 1977; Smith 1999). Burma's many nationalities never considered themselves to be assimilated into the polity of the independent Burma. While it is often claimed that Burma is one of the most ethnically diverse nations in the world (International Crisis Group 2003), there are no clear population figures, as the last comprehensive population census as noted earlier was conducted by the British in 1931. Both the military regimes and the armed opposition groups make their own estimates of population, so these figures are to be accepted with caution. Over 130 distinct ethnic groups based on linguistic, religious, and regional divisions are reported (Hla Tun Aung 2003; South 2003; Smith 1999). While the Burmans are clearly the majority, political fragmentation of the state remains a constant threat to military rule.

Independent democratic government lasted for only 14 years. Military power was severely challenged in 1988. The economy had so deteriorated that nation wide demonstrations were held against the regime that had ruled the country since 1962 (Lintner 1989). Burma as ruled by a series of army generals brought almost all of political, social, economic life under strict military control. At the end of General Ne Win's long rule (1962–1988), a country wide popular uprising against military rule, led by student activists, was brutally attacked by the army. Senior ranking *Tatmadaw* officials formed the State Law and Order Restoration Council (SLORC) in response to popular discontent. A brutal crackdown on protestors followed with an estimated 10,000 people killed by security forces (Smith 1999, 1).

The generals held an election in 1990, anticipating a return to a military run parliament, but their chosen party failed to win votes. Despite continuing arrests and severe restrictions on freedom of assembly and speech, the National League for Democracy (NLD) under the leadership of Aung San Suu Kyi won a landslide victory in the 1990 elections. Throughout the campaign, Aung San Suu Kyi had been kept under house arrest at her home in Rangoon, where she remains, at the time of this writing. When the National League for Democracy (NLD) won the majority of votes, the State Law and Order Restoration Council (SLORC) generals refused to give power to elected parliamentarians, insisting that a new constitution would first have to be drawn up by the military. The NLD and Aung San Suu Kyi effectively excluded from the constitutional process have been subjected to years of repressive actions including jail. Power was never transferred to the elected members of parliament. Martial law was declared and hundreds of people protesting military actions were shot. Many of the pro-democracy movement leaders, elected members of parliament, and supporters were imprisoned.

Notably after the military coup in 1988, the number of prisoners dramatically increased. Before 1988, there were about 40,000 people detained in Burmese prisons. After 1988, the number rose to about 60,000 with an additional 20,000 in labour camps for a total of 80,000 prisoners. Of this entire prison population, over 2,500 were political prisoners (AAPPB 2001; 2005).The military remains in power changing its name in 1997 to the State Peace and Development Council (SPDC).

The current State Peace and Development Council (SPDC) appoints the Prime Minister, senior officials, and determines national policy. The twelve members of the SPDC all military officers hold the most senior roles in army hierarchy. Cabinet ministers, except for two, are military officers with no background in governance. The SPDC does not meet on its own but joins once monthly cabinet meetings and thrice yearly meetings of all senior cabinet and military officials at which broad policy and strategic issues are decided. The Vice Senior General chairs the Trade Policy Committee, which meets once a week to rule on all decisions affecting external and internal economic regulation. Senior General Than Shwe chairs the Special Project Implementation Committee and Special Border Projects Committee which approves all decisions on infrastructure construction (e.g. bridges, dams, irrigation), energy projects and agricultural policy. Committees do not seek advice. People dare not offer it. Top generals frame the issues and make decisions based on limited knowledge.

As outlined earlier, the Burman administrative divisions are in seven primarily Burman ethnic Divisions (*taing*): Irrawaddy, Pegu, Magwe, Mandalay, Rangoon, Sagaing, and Tenasserim and seven ethnic States (*pyi nae*): Chin, Kachin, Karen, Karenni, Mon, Arakan, and Shan. Each State or Division is administered through a State/Division Peace and Development Council. These Councils are headed by an Area Commander supported by the Head of Department of General Administration and Police. The Ministry of Agriculture and Irrigation (MOAI) is represented by divisional or state managers of the Myanmar Agricultural Service (MAS) under the Settlement and Land Records Department (SLRD).

The 64 districts, the next administrative units, are managed like divisions, with a Peace and Development Council, a District Commander and a Deputy Commissioner of General Administration and Police. The MOAI is represented through the same agricultural groups. Beneath the district is the township, a total of 324, with about five per district. Each Township is managed by a Peace and Development Council chaired by the Head of the Department of General Administration. Both the MAS and the SLRD have township managers. Each township comprised a number of village tracts/groups (wards in urban areas) depending upon the population density. Each village group has a Chairman and a Peace and Development Council. The smallest administrative unit is the village, usually three to seven per group.

A military footprint of Burma overlaps the geographic map of States and Divisions. Fifty years (1948–1998) of construction and expansion of Burma's military, *Tatmadaw*, is explored in detail by Maung Aung Myoe (1999). The military regime has created thirteen military commands, sub-commands and an extensive surveillance network of military intelligence, detention, interrogation centres, prisons and prison labour camps. At the national, regional and local levels the regime attempts to quell dissent and boasts of 'military operations launched by the *Tatmadaw* at the cost of its lives, blood and sweat for turning the nation into a peaceful, modern and developed State as well as for ensuring its safety' with the function of the Southern Command to 'annihilate the remnants of Karen National Union insurgents' (Ministry of Information 2004, 16). Mary Callahan (2003, 221) argues that the relationship in Burma between the ruled and the rulers has long been one mediated by 'profound distrust and the constant threat of violence' and coercion. The military regime extols their successes of bringing together through violent means and of negotiating with armed opposition groups in returning to the 'legal fold'. But these agreements were between military leaders and the leaders of armed groups with no need to return their arms. All people living in areas where there are armed opposition forces are viewed as enemies.

The depth and extent of human rights abuses committed by the military on people of ethnic minorities living in border areas of Burma, including dispossession from land, forced evictions from villages, confiscation of labour and materials has been well documented. An example: *Dying Alive: An Investigation and Legal Assessment of Human Rights Violations Inflicted in Burma, with Particular Reference to the Internally Displaced, Eastern Peoples* (Horton 2005). Research conducted by David Arnott (*Forced Migration/Internal Displacement in Burma with an Emphasis on Government-Controlled Areas*, 2007) revealed reasons of forced migration from Irrawaddy, Magwe, Mandalay, West Pegu, Rangoon and Sagaing Divisions, and Kachin, Chin, Arakan, northern and eastern Shan State. A survey conducted with 560 refugees and migrants from Burma now living in India, Thailand, and Malaysia, revealed the following reasons for leaving their homes: land confiscation (39.1 per cent); food insecurity (69.8 per cent); forced labour (59.9 per cent); extortion/heavy and arbitrary taxation (60 per cent) and; ruinous agricultural cropping and marketing policies (18.6 per cent). These reasons for forced migration are closely connected to the lack of secure land tenure in Burma. Despite large numbers of people forced to flee, the majority remain inside Burma living under increasingly severe conditions.

Karenni State: Case Study

Research conducted in Karenni State (Hudson-Rodd and Sein Htay 2008) reveals the extent of military occupation of border areas. Karenni State lies in the eastern part of Burma bordering Shan and Karen States and an international boundary with Thailand. With an area of 11,7333 square kilometres, Karenni is the smallest of the seven States and the second smallest among the 14 States and Divisions. There are two Districts (Loikaw and Bawlakhe) with seven smaller townships (Hla Tun Aung 2003, 636). According to the Regime's Ministry of Information (2007, 40), Kayah State with a 'population of 318,700 is a microcosm of the

Union, represented by all the "national races" including Kayah, Kayan, Mono, Kayaw, Yintale, Gekho, Geba, Kachin, Kayin, Chin, Pa-O, Bamar, Mon, Rakhine, Shan, and Intha'. Karenni State's occupied land (net sown area plus fallow land) consists of nearly 50,000 acres each of *Le* land and *Ya* land, about 400 acres of Garden land and over 14,000 acres of *Taungya* land[3] (Hla Tun Aung 2003, 639). The average household size is 7.8, with over half (57.57 per cent) of the people living on less than five acres, and 10 per cent of the rural households are landless (FAO and World Bank 2004). According to the Central Statistical Organisation (CSO), *Report of 1997 Household Income and Expenditure Survey* (Yangon 1999) households in Karenni State have the lowest average monthly incomes in the nation adequate to meet only 41 per cent of their daily expenditures for example on food, housing, education, travel, fuel, light, and other basic necessities.

Karenni State is highly militarised (Bamforth, Lanjouw, Mortimer 2000). Twenty-eight SPDC regime army battalions, operating in Karenni State have displaced over 70,000 people, forcing them into new villages, rebuilt villages, relocation sites, or into hiding sites. Burma army units occupy and patrol areas of infrastructure development in southwest Karenni State, for re-building the road from Mawchi (Karenni State) to Toungoo (Karen State) and the Mawchi mines. Karenni resistance groups are attempting to stop the road construction as the road would facilitate army units travelling freely within Kayah State. The Burma army also uses two other armed groups, the Karenni National Peoples Liberation Front and the Karenni National Solidarity Organisation to control and attack the civilian population (Free Burma Rangers 2007).

Khun Mak Ko Ban, elected Member of Parliament (MP) for Phekon and Kayah representative for Democratic Organization for Kayan National Unity (DOKNU), now living in exile, submitted the following information concerning confiscation of residences and farm land in the Phekon and Moebye areas. The details of the military regime uncompensated confiscation of farmland, residences, and crops, were gathered by people inside Burma at considerable personal risk (May 2007). The following two maps (see Figures 9.7 and 9.8) illustrate the location of confiscation, due to military establishment of a base in the two towns. Despite an estimated 50 per cent of the national budget allotted to building up the armed forces, there is not enough to support the field units. In 1998, the SPDC in Rangoon, informed its field units that rations would be reduced and that they needed to 'fend for themselves' by producing their own food or getting it from villagers (Human Rights Documentation Unit 2000, 299). Since that order was delivered, villagers' land has been confiscated for army crops, food is taken with no compensation and villagers are forced to labour in their former fields, planting and harvesting crops for the army (see Table 9.1).

---

3   *Le* land refers to paddy land. *Ya* land refers to non-paddy land where pulses, beans, sesame and other short term cash crops are grown. Garden land is where flowers, fruits, vegetables and seasonal cash crops are grown. *Taungya* land is hill sloped land where shifting agriculture is done.

Findings in our research reveal the confiscated residential and farm lands of the residents of Phekon and Mobye Towns was given to retiring army personnel and to their relatives who were invited to move into the Township. The original ethnic Karenni residents reported feeling strongly that the army personnel, all of Burman (Bamar) ethnicity, were acting as if they were a 'master class'. The residents are

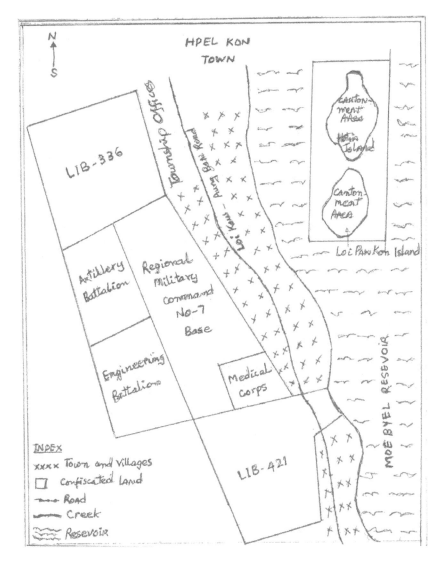

Figure 9.7   Phekon Town and Villages, Karenni/Kayah State

Source: Hand-drawn map used with permission from Burmese activist who wishes to remain anonymous. *Inside Researcher*, 29 May 2007, Phekon.

worried that these sentiments may erupt into 'racial' clashes in the future. The residents of both towns stated that they were being 'ethnically cleansed' and over 200 youth joined the Ka La La Ta, Kayah New Land, Ya La La Hpa Group, and the armed Karenni National Progressive Party (KNPP) at the time of 'peace' negotiations. The Ministry of SPDC Information (2003) states that the Kayah State Special Region-1 was a national armed group which 'returned to the legal fold' in 27 February 1992 in Mobye and Phekon Towns. There is no peace in the towns as army battalions, and retired army personnel live on confiscated Karenni lands.

Mobye Dam was built to the east of Phekon. All lands to the west of the dam were confiscated by the army. The people whose land was confiscated no longer had residences and there was no land remaining for them to build new houses. Phekon is about three miles long and about four phalon wide. The land confiscated by the army is five times bigger than the town area. The land was confiscated by the following army division: Light Infantry Battalion (LIB) 336, Artillery Battalion, Engineering Battalion, Regional Military Command Base, Medical Corps, and Light Infantry Battalion 421. Forty one residential houses and land they occupied were confiscated. At the time of confiscation the value of these residences were 1,460 Lakhs or 146,000,000 Kyats.[4] No compensation was given. Farmland of two hundred and eighteen owners, which included 1,346 household members, was confiscated. A total of 2,040 acres of farmland was taken. The land value at the time of confiscation was 448,500,000 Kyats. Crops worth 213,400,000 Kyats were lost with a total 2,121,900,000 Kyat loss of land and crops to the families. The value of the farm land has increased between 10 to 100 times since confiscation.

Due to construction of Mobye Dam, over 80 houses from Loi Pan Son Island and over 50 houses from Ho Tin Island were confiscated (See Figure 9.8). These people depended upon fishing for their survival and when their two villages were confiscated to become army land, there was great hardship for the residents. Around Mobye, because the army had taken the land there was no more space to expand the residential area. The army land is now three times bigger than the civilian town area as the Light Infantry Battalion 422 expands. One hundred and eight owners with 1,136 family members had their farmland confiscated. One thousand five hundred and seven (1,507) acres of land valued at the time of confiscation of 253,000,000 Kyat were confiscated. The value of the crops lost was 228,000,000 Kyat making a total loss to farming families of 480,000,000 Kyat.

---

4   There are two different exchange rates for the kyat: the official and the blackmarket with the former cited one US dollar as equivalent to K 6.69010 and the blackmarket (as of March 2008) of one US dollar to K 1100. (Source: http://en.wikipedia.org/wiki/Myanma_kyat).

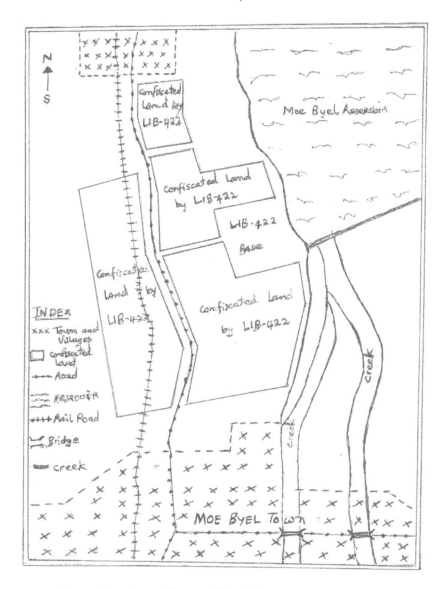

Figure 9.8   Mobye Town, Karenni/Kayah State

Source: Hand-drawn map used with permission from Burmese activist who wishes to remain anonymous. Inside Researcher, Phekon, 29 May 2007.

Table 9.1   State Security Network in Burma

State Security Network in Burma	
NATIONAL State Peace and Development Council (SPDC) Army Navy Air Force Office of Strategic Services (OSS) Directorate of Defence Intelligence Services (DDIS) National Intelligence Bureau (NIB)	
REGIONAL 13 Regional Commands Western Command (Arakan and Chin States) Central Command (Mandalay Division) Rangoon Command (Rangoon Division) Northern Command (Kachin State) Golden Triangle Command Coastal Command (Tenasserim Division) Southwestern Command (Irrawaddy Division) Southeastern Command (Mon, Karen States, Tenasserim Division) Eastern Command (Southern Shan State) Northeastern Command (Northern Shan State) Northwestern Command (Sagaing Division) Southern Command (Pegu and Magwe Division)	40 Sub Commands (Tactical Operations Commands) 360 Infantry Battalions (IB) (30 Battalions per Command) Light Infantry Divisions (LID) 77LID (est. 1966, Headquarters at Pegu) 88LID (est. 1967, Headquarters at Magwe) 99LID (est. 1968, Headquarters at Mektila) 66LID (mid-1970s, Headquarters at Prome) 55LID (est. late-1970s, Headquarters at Aungban) 44LID (est. late-1970s, Headquarters at Thaton) 33LID (est. mid-1980s, Headquarters at Sagaing) 22LID (est. 1987, Headquarters at Pa-an) 11LID (est. 1988, Headquarters at Hlegu) 101LID (est. 1991, Headquarters at Pakokku)
LOCAL Light Infantry Battalions (LIB) Infantry Battalions (IB) 20 Primary Detention Centres (Insein Prison and its annex: Yankin Township Military Registration Camp; Sanchaung Police Station; Mingladon DDSI Interrogation Centre; Kyeikkasan old race field; Hmawbe Ye Nyunt Training Camp (all in Rangoon Division); Bassein Township MIS Office (Irrawaddy Division); Special Branch 11 Office, Police Station No. 8 and Mandalay Prison (Mandalay Division); Tharawaddy and Pegu Township Prison (Pegu Division); Regiment 31 Headquarters (Thanbyuzayat Township); Moulmein MIS Office (Mon State); MI-5 Office (Pa-an Township, Karen State); Regiment 59 Headquarters (Maw Chee Township, Kayah state);	

Table 9.1 continued   State Security Network in Burma

State Security Network in Burma	
LOCAL continued Myitkyina Central Prison, No. 8, Northern Command Army Headquarters; Special Branch 11 Office (Kachin State). 35 Prisons Military Intelligence Service (MIS) Units 30 Military Intelligence Companies Military Interrogation Centres Paramilitary People's Forces	

Sources: Assistance Association for Political Prisoners (AAPP) Burma (2005), *The Darkness We See: Torture in Burma's Interrogation Centers and Prisons* (Mae Sot: AAPPB); Ball, D. (1998), 'The Burmese High Command and the Intelligence and Security Establishment', in D. Ball, *Burma's Military Secrets: Signals Intelligence (SIGINT) from 1941 to Cyber Warfare* (Bangkok: White Lotus); Bo Kyi (June 2000), 'The ICRC in Burma', *Irrawaddy* (Bangkok); Hudson-Rodd, N. (2004), based on displays, Defense Services, Museum (Rangoon); Human Rights Documentation Unit (HRDU), *Human Rights Yearbooks Burma* (1997, 1998, 1999, 2000, 2001), (National Council of the Government of the Union of Burma (NCGUB): Bangkok); Ministry of Information (2004), *Magnificent Myanmar 1988–2003* (Yangon: SPDC); Selth, A. (June 1997), *Burma's Intelligence Apparatus*, Working Paper No. 308 (Canberra: Australian National University).

Military Regime Administration

The Human Security Centre (2005, 26–7) determined that Burma was the country which had experienced the most 'conflict-years' between 1946 and 2003. A conflict-year refers to a calendar year in which a country has been involved in a state-based armed conflict of any type. The Human Security Centre counted years from the country's date of independence or 1946 whichever came first. Conflict types included intra-state, inter-state, colonial and internationalised intra-state conflicts. Because a country can experience several different armed conflicts at the same time, it can experience two or more conflict years within a single calendar year. Burma embroiled in six different intra-state conflicts in the 1990s, was the world's most conflict prone country with 232 conflict years. India came second with 156 conflict years. If only the measure for conflict had been determined by the number of years a state was involved in any conflict between 1946 and 2003, Israel would have been the top of the list (Human Security Centre 2005, 27). The following table (Table 9.2) based on the SPDC regime's own documents shows the agreements between the regime and the armed opposition groups, ethnicity, and geographic location in Burma. The extent of violence for one brief period of time to achieve 'peace' is detailed from the perspective of the Ministry of Defence (Ministry of Information 2004, 17) in these terms:

> In launching military operations nationwide from November to December 1997, the Tatmadaw had 5,781 major and minor engagements with armed insurgent groups, with 47 officers and 901 other ranks sacrificing their lives for the country and 161 officers and 2,352 other ranks getting wounded. The Tatmadaw captured 3,925 of the enemy dead and 598 live together with 3881 various sorts of arms.

Many different armed state and ethnic insurgent groups still exist in Burma, including an estimated 45,000 armed non-state combatants (International Institute for Strategic Studies 2005, 430–31). All groups are involved in arms production and sales. Both the military regime and the armed opposition groups use a variety of weapons including landmines and the use of children as soldiers. Small weapons, cheap easy to smuggle, hide, steal, capture, and buy are widely spread throughout Burma. Police, security patrols, state and non-state armed groups, traffickers of drugs and people all use them. At least 17 non-state armed groups have been using landmines in Burma since 1999. The Myanmar Defense Products Industries, a state owned enterprise produces anti-personnel mines. The United Wa State Army produces mines at its arms factory formerly belonging to the Burma Communist party, which was set up with help from China. This factory is reported to make landmines, ammunition, and other items. The Karen National Liberation Army (KNLA), the Democratic Karen Buddhist Army, the Chin National Army, and the Karenni Army, all have built blast and fragmentation mines, booby-traps, and landmines with anti-handling devices. Armed opposition groups lift SPDC regime landmines from the ground, seize military stocks during attacks, and buy in the shadowy arms market (International Campaign to Ban Landmines 2004, 939–40). In addition to locally produced landmines, others have been purchased from Chinese, Russian, Italian, and America manufactures (International Campaign to Ban Landmines 2005, 938).

The Ruling State

The State rules by decree not bound by any constitutional rights for public trials or any other human rights. There is no independent judiciary, no meaningful rule of law, and lack of political, intellectual, and legal discourse required for establishing a system of Burmese law to direct Burma towards transition to 'legality'. Burma remains guided by martial law as defined in Declaration No. 1/90 to prevent 'the disintegration of the Union, the disintegration of national unity and the perpetuation of Sovereignty'. These foundational aims of the State were reiterated by Khin Maung Win, Deputy Minister for Foreign Affairs of the Union of Myanmar in the State Constitution (January 2004). The SPDC has subordinate Peace and Development Councils ruling by decree at the division, township, state, city, ward, and village levels. The SPDC appoints justices to the Supreme Court who in turn appoint lower court judges with the approval of the SPDC. The court system includes courts at the township, state, district, and national levels. These courts adjudicate cases under decrees promulgated by the SPDC that effectively

Table 9.2   Agreements between the SPDC and the armed opposition groups, ethnicity, and geographic location in Burma

	Name of Group	Location/Ethnicity	Date of Entry to 'Legal Fold'
1	Myanmar National Democracy Alliance (MNDA)	Northern Shan State Special Region 1 (Kokang)	31 March 1989
2	Myanmar National Solidarity Party (MNSP)	Northern Shan State Special Region 2 (Wa)	9 May 1989
3	National Democracy Alliance Army (NDAA)	Eastern Shan State Special Region 4 (Shan/Akha)	30 June 1989
4	Shan State Army (SSA)	Northern Shan State Special Region 3 (Shan)	24 September 1989
5	New Democratic Army (NDA)	Kachin State Special Region 1	15 December 1989
6	Kachin Defence Army (KDA)	Northern Shan State Special Region 5	11 January 1991
7	Pa-O National Organization (PNO)	Southern Shan State Special Region 6 (Pa-O)	18 February 1991
8	Palaung State Liberation Party (PSLP)	Northern Shan State Special Region 7 (Palaung)	21 April 1991
9	Kayan National Guard (KNG)	Kayah State Special Region 1	27 February 1992
10	Kachin Independence Organization (KIO)	Kachin State Special Region 2	24 February 1994
11	Kayinni National People's Liberation Front (KNPLF)	Kayah State Special Region 2	9 May 1994
12	Kayah New Land Party (KNLP)	Kayah State Special Region 3	26 July 1994
13	Shan State Nationalities People's Liberation Organization (SNPLO)	Shan State (Pa-O)	9 October 1994
14	Kayinni National Progressive Party (KNPP)	Kayah State (Kayinni)	21 March 1995
15	New Mon State Party (NMSP)	Mon State	29 June 1996
16	Mon Tai Army (MTA)	Mon State	5 January 1996
17	Burma Communist Party (BCP)	Rakhine State	6 April 1997

Sources: Adapted from Khin Maung Win (January 2004), Deputy Minister for Foreign Affairs, Union of Myanmar, (19–20) and Ministry of Information (2004) SPDC (15).

hold the force of the law 'what the generals from day to day decide it to be' (Gutter and Sen 2001, 14). Laws are applied selectively and arbitrarily by the military regime seeking to maintain control (Steinberg 1995, 7).

Almost all of the criminal, civil, corporate, and commercial laws stem from British rule. The latest law regarding the judiciary is the Judiciary Law promulgated 27 June 2000 (amended since). Based on that law, the SPDC constituted the Supreme Court (1 Chief Justice, 2 Deputy Chief Justices, and a minimum of 7 judges to a maximum of 12 judges) with military personnel. The Supreme Court sits in Rangoon and Mandalay and if necessary in any other appropriate place. Under the Supreme Courts are the Chief Court, the State or Divisional Courts, the District Courts and Township Courts, with powers of civil and criminal jurisdiction. The judiciary is not independent. It has to obey SPDC orders in adjudicating cases. The SPDC appoints only those men loyal to the regime to be Chief Court Judges. Therefore the Chief Justice and the Attorney General comply with all decrees of the SPDC.

The laws, courts, and other legal structures have been manipulated over decades by the regime to punish perceived enemies and to harass and intimidate civilian population. While the number varies of people held as prisoners of conscience and political prisoners, about 1500 political prisoners are held at any one time. Outcomes of trials concerning political activists or any person seen to be a critic of the regime are determined before trial. 'Evidence' is made up or taken after individuals are picked up and tortured in detention centres. Only the arguments which lead to a guilty verdict are allowed by the judge. Lawyers cannot defend their clients independently, especially in cases where there is political or state interest. Decisions are predetermined by the SPDC. The special hostility held by the SPDC towards lawyers may result from the regime's view that lawyers initiated and supported the large demonstrations against the military regime in 1975 and in 1988 (US Department of State 2002, 22).

With the remnants of the British era legal system remain formally in place, the court system and its operations are severely flawed especially for handling political cases. The military junta controls the three key powers of the State: the judiciary, the executive, and the legislative. Laws such as the Emergency Provisions Act, the Unlawful Associations Act, the Habitual Offenders Act, and the Law Safeguarding the State from the Danger of Subversive Elements are used extensively to stifle any dissent or questioning of the regime with political trials not open to the public. Human rights and democracy activists, people participating in peaceful demonstrations, speaking, writing, singing any words deemed to be against the regime, have been arrested, tortured, and jailed in the many prisons located in Burma. Most are given long prison terms, in isolation cells, and placed in remote prisons, making family visits if allowed, almost impossible (Hudson-Rodd and Myo Nyunt 2005). Political prisoners and prisoners of conscience are sometimes charged under a section of the criminal act which authorises them being sent to hard labour camps (AAPPB 2001, 1). For example, Aye Tha Aung, Secretary of the Committee Representing the People's Parliament was accused of violating publication and emergency laws and given a 21-year prison term (7 June 2000). He was refused any legal defense and sent to a hard labour prison camp where he with other prisoners was forced to construct the Myitkyina airport. He was then transferred to another prison labour camp in Sumprabum to build a highway. Both labour prison camps are in northern Burma, Kachin State. In 2007, the regime continued to rule by decree and was not bound by any constitutional provisions providing for fair public trials or any other rights. 'Pervasive corruption further served to undermine the impartiality of the justice system' (US Department of State 2007, 3).

Conclusion

David Steinberg (2005) an historian of Burma who has met members of the military leadership, confirms that the current regime see themselves as the sole guardians of the country's sovereignty and regard any outside interference with hostility especially the support or encouragement of ethnic rebellions against the Burmese leaders. Ethnicity and violence against those people from a non-Burman background has been emphasised by international groups, yet violence has been and continues to be perpetuated against all men, women, and children who seek to survive in Burma regardless of ethnicity. Aung San, the independence hero addressed ethnic minority leaders at the historic Panglong Conference where ethnic principles of the future Union of Burma were agreed:

> If we want the nation to prosper, we must pool our resources, manpower, wealth, skills, and work together. If we are divided, the Karens, the Shans, the Kachins, the Chins, the Burmese, the Mons and the Arakanese, each pulling in a different direction, the Union will be torn, and we will all come to grief. Let us unite and work together (Aung San quoted in Maung Maung 1962, 124).

Within six months Aung San and most men of his cabinet had been assassinated. This dramatic loss of life and conflict has dominated in Burma from independence to now with ethnic groups seeking recognition based on their separate identity, language, religion, difference. As Amartya Sen explores in his critique of ethnicity and violence, our shared humanity is severely challenged when our differences are narrowed into one system of categorisation. The illusion of unique identity is much more divisive than the universe of plurality and diversity which characterise the worlds in which people of Burma live: 'Violence is fomented by the imposition of singular and belligerent identities on gullible people, championed by proficient artisans of terror' (Sen 2006, 2). The military regime in Burma seeks to become more proficient in their acts of terror and surveillance. The enhanced focus upon ethnic diversity and ethnic rights in Burma by journalists, academics, international organisations may have brought more awareness to these groups, but it also aggravates and accentuates the problems of classification of people according to certain criteria of language, religion, background. The ethnic diversity of Burma is more complex than major ethnic categories declare. The ethnic groups and the military regime both seek to represent an authentic version of themselves. These groups are engaged in struggle to impose their own versions of reality, to confirm their identity and cultural legacy. Ethnicity is used as a screen for other political interests. The SPDC regime benefits from this struggle by maintaining control over a divided nation. Aung San Suu Kyi (1995, 185) leads quietly for peaceful resolution to decades of illegitimate military rule by force and conflict through principles of justice, compassion, and truth.

> It is man's vision of a world fit for rational, civilised humanity which leads him to dare and to suffer to build societies free from want and fear. Concepts such as truth, justice, and compassion cannot be dismissed as trite when these are often the only bulwarks which stand against ruthless power.

Postscript: The Cyclone Nagris 2008

On 2 May, tropical Cyclone Nargis stormed across the Bay of Bengal, hit the Irrawaddy delta cutting a path of destruction direct to Rangoon. Winds exceeding 190 kilometres per hour flattened houses, uprooted trees, brought down power lines, disrupted water supplies, tore apart more sturdy Rangoon buildings, destroyed port facilities and boats. The storm surge submerged over 5,000 square kilometres of land in the delta, flattened, flooded, totally destroyed entire towns with Bogale, Labutta and Pantanaw particularly hit hard.

The State Peace and Development Council (SPDC) declared a state of emergency across five regions: Rangoon, Irrawaddy, and Pegu Divisions and Karen and Mon States, home to over 24 million people. Only 350 people died according to the regime, while the United Nations and foreign diplomats estimated at least 25,000 people were dead and one million homeless. Save the Children Fund, one of the few aid agencies permitted to operate inside Burma, claimed that nearly half of the dead and missing were children. The Food and Agriculture Organisation predicted dire food shortages as 65 per cent of Burma's rice, 80 per cent of fish and prawn aquaculture, 50 per cent of poultry and 40 per cent of pigs were grown in the affected region. Rice fields, oil palm, rubber plantations, most of rice storage silos were destroyed.

Political oppression and absence of freedom has magnified the suffering of the people in Burma. Despite knowledge of the impending disaster, the SPDC which controls all media, refused to warn people. The Indian Meteorological Department issued the SPDC forty-one general warnings of an impending cyclone, from 30 April giving hourly updates on the path, speed, severity, and locations where the cyclone would hit land. The Asian Disaster Preparedness Centre in Bangkok also issued cyclone warnings to SPDC authorities. The regime's response has been callous. Immediate assistance to cyclone victims was provided by hundreds of monks in Rangoon, who cleared storm debris and provided food and shelter. The regime watched. Despite massive immediate aid needed, the regime refused aid workers already inside Burma free movement. Food, shelter, clean drinking water, mosquito nets, and medical supplies are critical needs. Dead bodies and flooding contaminated water sources. People have lost all water containers. Sanitation is non-existent. All face starvation, diseases, and death, lacking food, water and sanitation. Tens of thousands of survivors are crammed into monasteries, schools and other buildings in towns which were at bare survival level before the cyclone. Limited supplies of bottled water, high energy biscuits, rice have been unloaded at Rangoon airport. Much of these items have gone directly to the regime, or sold at inflated rates to victims of the cyclone.

The natural disaster occurred just before the SPDC referendum on 10 May to adopt a new constitution which would ensure continuation of military rule. Despite the humanitarian crisis the regime held the referendum in all but the most affected southern regions.

UN agencies, international NGOs and many countries have offered aid. Yet the regime thwarts this comprehensive assistance needed to bring relief and hope to the people. How far should the international community go in challenging the right to national sovereignty when a government denies its most basic responsibility to protect citizens faced with mass suffering and loss of life during a natural disaster? France invoked the concept of 'responsibility to protect (R2P)' as a basis for Security Council resolution to authorise relief efforts. Under this principle when a national government refuses to protect its own people the international community under the auspices of the Security Council must assume this role. Some UN members argue this would stretch the limited scope of the concept covering only protection of people from genocide, war crimes, ethnic cleansing, and crimes against humanity.

Should people be denied external help because they are hostage to an oppressive regime? The application of R2P to the crisis in Burma would strongly demonstrate, especially to Asian countries, the importance and viability of this international norm. I conducted research (2007–2008) on the plight of farmers in the now devastated areas. The farmers and their lawyers were using all means possible to contest illegal, arbitrary confiscation of their land by the SPDC. Farmers and their representatives wanted the world to know that they were doing their utmost to achieve rule of law and their rights to survive. We need more than ever to recognise their human dignity and support them by respecting our international obligations to human rights. Our lives are diminished by lack of support to the people of Burma.

Selected References

Amnesty International (AI) Myanmar (1996), *Human Rights Violations Against Ethnic Minorities* (London: AI).

Amnesty International (AI) Myanmar (1997), *Myanmar: Ethnic Minority Rights Under Attack* (London: AI).

Amnesty International (AI) Myanmar (2000), *Exodus from the Shan State* (London: AI).

Arnott, David (16 April 2007), *Forced Migration/Internal Displacement in Burma, with an Emphasis on Government Controlled Areas* (draft copy).

Asia Development Bank (ADB) (2006), *Millennium Development Goals: Progress in Asia and the Pacific 2006* (Manilla: ADB).

Asian Human Rights Commission (AHRC) (1997), *Voice of the Hungry Nation* (Hong Kong: AHRC).

Assistance Association for Political Prisoners Burma (AAPPB) (2005), *The Darkness we See: Torture in Burma's Interrogation Centers and Prisons* (Mae Sot: AAPPB).

Aung San Suu Kyi (1995), *Freedom From Fear and Other Writings* (Harmondsworth: Penguin).

Bamforth, V; Lanjouw, S. and G. Mortimer (2000), *Conflict and Displacement in Karenni: The Need for Considered* Responses (Chiang Mai: Burma Ethnic Research Group (BERG)).

Burma's Economic Watch (29 March 2006), 'Burma's economic prospects: Testimony before the Senate Foreign Relations Subcommittee on East Asian and Pacific Affairs' (Burma's Economic Watch).

Burma Ethnic Research Group (BERG) (1998), *Forgotten Victims of a Hidden War: Internally Displaced Karen in Burma* (Chiang Mai: Nopburee Press).

Callahan, Mary (2003), *Making Enemies: War and State Building in Burma* (Ithaca: Cornell University Press).

Central Statistical Organization (CSO) (1999), *Report of 1997 Household and Expenditure Survey* (Yangon: CSO).

Council on Foreign Relations (June 2003), *Burma: Time for a Change* (New York).

Frontier Areas Committee of Enquiry 1947 (24 April 1947), *Report Presented to His Majesty's Government in the United Kingdom and the Government of Burma (Part 1: Report)* (Maymyo, Rangoon: Superintendent Government Printing and Stationery, Burma).

Gutter, P and Sen, B.K. (December 2001), *Burma's State Protection Law: An Analysis of the Broadest Law in the World* (Bangkok: Burma Lawyer Council).

Hla Tun Aung (May 2003), *Myanmar: The Study of Processes and Patterns* (National Yangon: Centre for Human Resource Development, Ministry of Eduction, Myanmar).

Horton, Guy (2005), *Dying Alive: An Investigation and Legal Assessment of Human Rights Violations Inflicted in Burma, with Particular Reference to the Internally Displaced, Eastern Peoples* (Co-funded by The Netherlands Ministry for Development Co-operation, Changmai: Images Asia).

Houtman, G. (1999), *Mental Culture in Burmese Crisis Politics* (Tokyo: Institute for the Study of Languages and Cultures of Asia and Africa, Tokyo University of Foreign Studies).

Hudson-Rodd, N., Myo Nyunt, Saw Thamain Tun, and Sein, Htay (June 2004), *State Induced Violence and Poverty in Burma, Research Report submitted to ILO Geneva on behalf of Federation of Trade Unions-Burma* (FTUB).

Hudson-Rodd, N. and Myo Nyunt (2005), 'The Military Occupation of Burma', *Geopolitics* 10, 500–521.

Hudson-Rodd and Sein, Htay (2008), 'Arbitrary Confiscation of Farmers' Land by the State Peace and Development Council (SPDC) Military Regime in Burma' (Washington: The Burma Fund).

Human Rights Documentation Unit (HRDU) (2000), *Human Rights Yearbook Burma (Myanmar)* (Washington, D.C. and Nonthaburi, Thailand: NCGUB).

Human Security Centre (2005), *Human Security Report 2005: War and Peace in the 21st Century* (Human Security Centre, University of British Columbia, New York and Oxford: Oxford University Press).

International Crisis Group (ICG) (7 May 2003), *Myanmar Backgrounder: Ethnic Minority Politics*, ICG Asia Report No. 52 (Brussels: ICG).

International Institute for Strategic Studies (2005), *The Military Balance 2005–2006*, (London: Routledge).

International Labour Organization (ILO) (2006), *Individual Observation Concerning Forced Labour Convention, 1930 (No. 29) Myanmar (ratification 1955)*, Document No. (ilolex) 062006MMR029.

Karen Women's Organization (KWO) (February 2007), *State of Terror: The Ongoing Rape, Murder, Torture and Forced labour Suffered by Women Living Under the Burmese Military Regime in Karen State* (Mae Sot: KWO South).

Khaing, Soe Naing Aung (2000), *Brief History of the National Democratic Movement of Ethnic Nationalities: Union of Burma* (Bangkok: National Democratic Front).

Khin, Maung Win, Deputy Minister for Foreign Affairs, Union of Myanmar (January 2004), *Myanmar Roadmap to Democracy: The Way Forward,* Seminar on Understanding Myanmar, Myanmar Institute of Strategic and International Studies (MICT Park, Yangon 27–28 January).

Lanser, T. (2005), 'Media and human rights', in R. Smith and C. van den Anker (eds), *The Essentials of Human Rights* (New York: Oxford University Press).

Lintner, B. (1989), *Outrage: Burma's Struggle for Independence* (Hong Kong: Review Publishing).

Lokethar Pyithu Neizen (26 September 1990).

Maung, Aung Myoe (1999), *Building the Tatmadaw: The Organizational Development of the Armed Forces in Myanmar 1948–1998* (Canberra: Australian National University).

Maung, Maung (1962), *Aung San of Burma* (The Hague: Martinus Nijhoff).

Maung, Myint (2004), 'Constraints Faced by Researchers' (Unpublished Paper, Rangoon).

Maung, Myint (2007), 'Open letter to editors on regime denial of academic involvement in Burma policy-making sent to editors of BBC, RFA, DVB, Mizzima News, Irrawaddy News, Burma Net News, and Aljazeera' (personal communication).

Ministry of Information (2004), *Magnificent Myanmar (1988–2003)* (Yangon: Information and Public Relations Department, Union of Myanmar).

Ministry of Information (2007), *Chronicle of National Development: Comparison between Period Preceding 1988 and after (up to 31 December 2006)* (Yangon: Printing and Publishing Enterprise).

Myint, Thant (2001), *The Making of Modern Burma* (Cambridge: Cambridge University Press).

Narinjara News (2006), 'Burma stands second, Bangladesh third on corruption list', 7 November.

Selth, A. (2002), *Burma's Armed Forces: Power Without Glory* (Norwalk, CT: EastBridge).

Sen, Amartya (1999a), *Development as Freedom: Human Capability and Global Need* (New York: Anchor Press).

Sen, Amartya (1999b), 'Global justice, beyond international equity', in I. Kaul, I. Grunberg and M.A. Stern (eds), *Global Public Goods, International Cooperation in the 21st Century* (New York and London: UNDP and Oxford University Press).

Sen, Amartya (2006), *Identity and Violence: The Illusion of Destiny* (London: Allen Lane).

Shan Human Rights Foundation (SHRF) (1996), *Uprooting the Shan* (Chiang Mai: SHRF).

Silverstein, J. (1997) *Burma: Military Rule and Politics of Stagnation* (Ithaca: Cornel University Press).

Silverstein, J. (October 1996a), 'Burma's Uneven Struggle', *Journal of Democracy* 7: 4, 88–102.

Smith, M. (1999), *Burma: Insurgency and the Politics of Ethnicity* (London: Zed Books).

South, Ashley (February 2007), *Burma: The Changing Nature of Displacement Crises*, Refuge Studies Centre (RSC) Working Paper No. 39 (Oxford: Department of International Development, University of Oxford).

Steinberg, D. (1995), 'Civil Society in Burma', *The FDL Quarterly,* Winter.

Steinberg, D. (2005), 'Burma/Myanmar: The role of the military in the economy', *Burma Economic Watch*, vol. 1 (Burma Economic Watch).

The Economist (2004), *Economist Country Profile 2004: Myanmar (Burma).*

Tun, Aung Chan (July 2000), 'Historians and the Search for Myanmar Nationhood', Paper Presented at the *16th Conference of the International Association of Historians of Asia*, Kota Kinabalu, Malaysia.

United Nations General Assembly (UNGA) (21 September 2006), *Situation of Human Rights in Myanmar*, 61st Session, Agenda item 67 (c), A/61/369.

United Nations, Second Regular Session (11 to 13 September 2006), *Assistance to Myanmar, Note by the Administrator*, Executive Board of the UNDP and of the UNPF, DP/2006/43.

US Central Intelligence Agency (CIA) (19 December 2006), *World Fact Book*.

US Department of State (2002), *Country Commercial Guide 2002 Burma*, (Washington, D.C.).

WHO (2000), *World Health Organization Annual Report* (Geneva: WHO).

Yawnghwe, C.T. (10 December 2001), 'Burma and National Reconciliation: Ethnic Conflict and State-Society Dysfunction', *Legal Issues on Burma Journal* 10: 10.

Internet-based References

Amnesty International, *Myanmar Leaving Home*, AI Index: ASA 16/023/2005, 8 September 2005, <http://web.amnesty.org/library/index/engasa160232005> accessed 10 July 2007.

Amnesty International Report 2007, *Myanmar*, <http://thereport.amnesty.org/eng/Regions/Asia-Pacific/Myanmar> accessed 13 December.

Assistance Association for Political Prisoners Burma (AAPP), 8 March 2001, *A Glimpse of Political Prisons and Political Prison Conditions in Burma*, Mae Sot: AAPPB, <www.aappb.org/report3_%20Glimpse.pdf> accessed 1 February 2008.

FAO and World Bank, 2004, *Myanmar Agricultural Sector Review Investment Strategy*, Volume 1-Sector Review, <www.ibiblio.org/obl/docs4/FAO-ASR-MainReport2004.pdf> accessed 9 December 2007.

Free Burma Rangers, 7 November 2007, 'FBR Report: Southwest Karenni State', <www.freeburmarangers.org/Reports/2007/20071112.html>, accessed 14 December 2006.

Garrett, L., 2005, *HIV and National Security: Where are the Links? A Council on Foreign Relations Report*, New York: Council on Foreign Relations, <www.cfr.org/content/publications/attachments/HIV_National_Security.pdf> accessed 8 May 2008.

Heritage Foundation, 2008, 'Burma', *Index of Economic Freedom 2008*, <www.heritage.org/index/country.clm?id=Burma> accessed 29 February 2008.

Human Rights Watch, 2007, *Burma*, New York: Human Rights Watch, <http://hrw.org/englishwr2k7/docs/2007/01/11/burma14865.htm> accessed 10 August 2007.

International Campaign to Ban Landmines (October 2004), *Landmine Report 2004: Towards a Mine-Free World: Special Five Year Review*, <www.icbl.org/lm/2004> Accessed 20 August 2007.

International Campaign to Ban Landmines (October 2005), *Landmine Report 2004: Towards a Mine-Free World,* <www.icbl.org/lm/2005> Accessed 20 August 2007.

Inter-Parliamentary Union (IPU), *Myanmar,* Human Rights of Parliamentarians 166th Session of the Inter-Parliamentary Council, 6 May 2000, Amman, 1–4, IPU: Geneva, <www.ipu.org/hr-e/166/myn01.htm> accessed 8 May 2008.

Kyi May Kaung, 4 December 2007, 'Post-crisis economic fallout in Burma', <www.mizzimaNews/EdOp/2007/Dec/PostcrisiseconomicfalloutinBurma(Analysis).htmldecember> Accessed 20 December 2007.

NCGUB (The National Coalition Government of the Union of Burma), August 2007, 'Situation Update of Members of Parliament', <www.ncgub.net/staticpages/index.php/MP-update-August2007>, accessed 8 May 2008.

Silverstein, J., 1996b, 'Conclusion', in J. Silverstein, *Burma's Struggle for Democracy: The Army Against the People*, <http://epress.anu.edu.au/mdap/mobile_devices/ch05s09.html>, accessed 8 May 2008.

US Department of State (2007), *Doing Business in Burma: A Country Commercial Guide for US Companies*, online copy <www.state.gov/e/eeb/ifd/2007/80685.htm> accessed 22 January 2008.

Index